Vanier

The Structures and Properties of Solids

a series of student texts

General Editor:
Professor Bryan R. Coles

The Structures and Properties of Solids 3

The Dynamics of Atoms in Crystals

W. Cochran, F.R.S.,
Professor of Physics, University of Edinburgh

Edward Arnold

First published 1973 by Edward Arnold (Publishers) Limited
25 Hill Street, London W1X 8LL

Boards Edn. ISBN 0 7131 2438 5
Paper Edn. ISBN 0 7131 2439 3

Printed in Great Britain by
William Clowes & Sons Ltd., London, Colchester and Beccles

General Editor's Preface

Most of the solids with which physicists or materials scientists are concerned are crystalline. In consequence two central components of an understanding of solids are on the one hand an understanding of the types of static atomic arrangement and consequent crystal symmetry manifested by real solids, and on the other hand an understanding of the excitations that crystal lattices can experience and their consequences for the thermal, optical and electrical properties. This second component is the subject of Professor Cochran's book (the first having been provided in the previous book in this series *The Crystal Structure of Solids* by Drs. Brown and Forsyth), and he brings to it the experience of someone who has not only appreciated but also strongly influenced the new developments in and attitudes towards the dynamics of atoms in crystals.

He has seen developments in neutron scattering studies of crystals transform the dispersion curves and density of states functions for lattice vibrations from theoretical constructs to experimental observables which provide a detailed challenge to microscopical theories. Other technical developments still in progress are associated with the possibilities brought by lasers of expanding the Brillouin and Raman studies discussed in Chapter 7, and of applying the infrared spectroscopy of section 7.5 to augment other studies of crystal defects familiar to the reader of Dr. Henderson's book in this series *Defects in Crystalline Solids.*

The lattice dynamics of defective crystals is too large and complex a subject to be covered in the present book, but Professor Cochran has been able in his final chapter to show the strong links between lattice dynamics and phase transitions, a topic currently of intense theoretical and experimental interest.

BRC

Imperial College,
London,
1973.

Preface

This is intended to be an intermediate level text in that it assumes an acquaintance with topics which a student will meet in the first years of an undergraduate course, but makes no claim to lead him to the frontiers of knowledge of the subject. References, mainly to books and review articles, are given at the end of each chapter and will enable the reader to proceed further if he wishes. Some topics, such as the application of group theory to crystal dynamics, and the dynamics of imperfect crystals, have been excluded as being beyond the level at which I was aiming.

In writing this book, I lost confidence at one point that I was in fact maintaining the level within the correct limits, and without the enthusiastic advice of Dr. G. Dolling who read several chapters and made suggestions for improvements, it might never have been completed. I am also grateful to him for providing material which has been incorporated in Chapter 6.

Edinburgh
1973

WC

Contents

1

Introduction

This book is one of a group of monographs dealing with aspects of solid state physics. Other topics include, Crystal Structure, Electronic Structure, Magnetic Properties, and Defects, all in relation to crystalline solids. When one considers the major properties of solids which do not come under any of these headings, one finds that many of them can be accounted for by treating the crystal as a perfect one in which the atoms make small oscillatory displacements from their equilibrium positions. This topic, often called Lattice Dynamics, is the subject of this book. The theory of lattice dynamics makes possible a more unified treatment of various parts of solid state physics such as the theory of specific heat of crystals, of their optical and dielectric properties, of certain aspects of the interaction of radiations such as X-rays and neutrons with crystals, and of their thermal conductivity. These are branches of the subject which at first sight appear to have little connection with one another. Of course it is not possible to consider the dynamical properties of a crystal in complete isolation from other properties, structural, electronic or magnetic, and there must be some overlap with other books in this series. Some knowledge of crystal structures and of crystallography will be assumed, and a knowledge of at least the elementary aspects of electron energy band theory will be a help to the reader, but an effort has been made to make the book self contained even when this has meant a degree of overlap with other books of the series.

Solid state physics, like most things, had no definite beginning, but we now recognize three or four papers as marking the beginnings, if not the beginning. 'Planck's theory of radiation and the theory of specific heat' was published by A. Einstein in 1907, 'On vibrations in space lattices' by M. Born and T. von Karman in 1912, and 'On the theory of specific heat' by P. Debye in 1912. Einstein took the atoms of a crystal to be independent oscillators each having the same (circular) frequency ω_E and able to vibrate freely in space. He assumed, following Planck, that the energy of each oscillator was quantized in units of $\hbar\omega_E$, and showed that the average energy of a crystal of N atoms at temperature T would then be given by

$$\bar{U} = \frac{3N\hbar\omega_E}{\exp\left(\dfrac{\hbar\omega_E}{k_B T}\right) - 1}$$

where k_B is Boltzmann's constant. The specific heat is then given by

$$C_v = \frac{d\bar{U}}{dT}$$

C_v is shown as a function of T/θ_E in Fig. 1.1, where θ_E is a convenient abbreviation for $\hbar\omega_E/k_B$ and has the dimensions of temperature. C_v is zero at $T = 0$ and rises asymptotically to the value $3Nk_B = 3R$ when $T \gg \theta_E$. For high temperatures therefore, quantization is unimportant and the specific heat has the same value as if each degree of freedom of the system had energy $\frac{1}{2}k_B T$, whereas at lower temperatures there is a pronounced deviation from this result of the theorem of equipartition of energy. In the Einstein theory the dynamics of a crystal is characterized by one parameter, ω_E, or more conveniently by

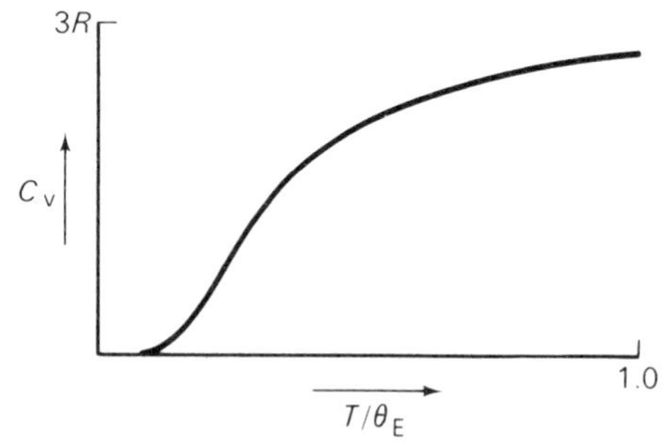

Figure 1.1 The specific heat as a function of reduced temperature, based on Einstein's model for the dynamics of a crystal.

the 'characteristic temperature' θ_E which is proportional to ω_E. In 'hard' crystals such as diamond the atoms vibrate with a high frequency, room temperature is a low temperature relative to θ_E and at ordinary temperatures $C_v < 3R$. In a 'soft' crystal, such as lead, ordinary temperatures are high compared with θ_E and $C_v \approx 3R$. Thus both Dulong and Petit's law for specific heats and the deviations from it are successfully accounted for.

Einstein's theory does not give quantitatively correct results for any material over a range of temperature because it is based on a model of the crystal which is much too simple. Born and von Karman used a more realistic model; they assumed the truth of the lattice theory of crystal structure, that the atoms are arranged in a periodic three-dimensional array. This was still a hypothesis in 1912; the discovery of the diffraction of X-rays by a crystal was published by Friedrich, Knipping and von Laue in 1913 and ranks as another 'founding paper' for solid state physics. In such an array the force on an atom depends not on its displacement form its equilibrium position but on its displacement relative to its neighbours, presumably its near neighbours, unless interatomic forces have a range large compared with the distance between nearest neighbour atoms. The motion of such a system turns out to be most readily described not in terms of the vibrations of individual atoms, but in terms of travelling waves, named lattice vibrations by Born, each characterized by a wave vector, a frequency and certain polarization properties. These waves are the normal modes of vibration of the system, and the energy of each is quantized in the same way as for a simple harmonic oscillator of the same frequency. Instead of one frequency associated with the crystal there is now a range of frequencies, or frequency distribution, which depends in a complicated way on the interatomic forces. Born and von Karman took this factor into account in their paper 'On the theory of specific heat' which they published in 1913. Debye's 1912 paper with the same title set out a theory which was less accurate in principle than that of Born and von Karman. It was, however, more successful in practice because of its simplicity. Debye's theory treated the normal modes of vibration as if they were waves in a continuous isotropic medium, instead of in a system in which the mass is concentrated at discrete points. This greatly simplifies the frequency distribution, and like Einstein's theory makes C_V the same function of T/θ_D for all crystals, where θ_D is the Debye characteristic temperature of a crystal. Since the frequency distribution introduced by Debye has some of the features of the actual frequency distribution of a crystal, the theory fits adequately all but the most accurate measurements of the specific heat of simple crystals. It was in fact not until the 1930's that the deficiencies of Debye's theory began to be noticed in comparisons of theory and experimental results, and the correct explanation was given in terms of Born and von Karman's theory, by Blackman.

Other early papers on lattice dynamics included papers by Debye and by Waller on the effect of the thermal motion of the atoms on the intensity of X-ray reflection from a crystal. Not only is the intensity of Bragg reflection diminished by the lattice vibrations, the latter produce definite features in the 'diffusely scattered' radiation which is seen in directions not allowed by Bragg's law. This was discovered experimentally by Laval in 1938, and correctly explained by him in terms of the Born von Karman theory.

If the force on an atom were precisely proportional to its displacement relative to neighbouring atoms each normal mode would propagate quite independently of all others, the principle of superposition would apply and we would have what is quaintly termed a harmonic crystal. Such a crystal would have no coefficient of expansion and its elastic properties would be independent of temperature for example, unlike real crystals in which the forces are only approximately linear. The theory of lattice dynamics is much more difficult when these 'anharmonic effects' are included. While their presence can often be neglected as a good approximation, they are particularly important for understanding thermal conductivity, or rather thermal resistance.

In a crystal in which the thermally excited modes of vibration superposed linearly there would be no mechanisms for scattering them in the absence of impurities and imperfections and therefore no thermal resistance. Anharmonic effects cause the waves in the crystal to scatter from one another. A basic step in the theory of thermal conductivity was made by Peierls in 1929. Peierls also contributed to the theory of scattering of electrons by lattice vibrations, which contribute to the electrical resistance of even a harmonic crystal.

Electro-magnetic waves in or near the optical region of the spectrum can interact with lattice vibrations, particularly in ionic crystals and such effects have become the basis of some experimental investigations of lattice vibrations with the advent of lasers. The basic theory of the more prominent effects was worked out mainly by Blackman and other collaborators of Born in the 1930's.

The dynamical theory of crystal lattices published by M. Born and K. Huang in 1954, is still the authoritative work on most aspects of the subject. The principal impetus in the past decade has come from the development of the technique of neutron inelastic scattering. Neutrons which have come into thermal equilibrium with matter at about room temperature have energies of the same order of magnitude as a quantum of lattice vibrational energy, for which the name phonon is used. The wavelength associated with a beam of neutrons all having the same energy, in this range, is of the same order of magnitude as interatomic distances, and the beam will be diffracted by a crystal. Most of the intensity is diffracted in accordance with Bragg's law. For this process the neutron energy or wavelength is not changed and one speaks of elastic scattering. The neutron beam is however diffracted in other directions by the thermally excited travelling waves in the crystal and exchanges energy with them in the process, in units of the phonon energy, which is directly proportional to frequency. Consequently by measuring the change in direction and in energy of the scattered neutrons it is possible to get more detailed information about the frequencies and polarization properties of the lattice vibrations than by any other experimental method. The information obtained in this way has caused renewed interest in the subject of lattice dynamics, which, as we have tried to show in this introduction, has a history longer than almost any other branch of solid state physics.

Suggestions for further reading

Born, M. and Huang, K. 1954, *The Dynamical Theory of Crystal Lattices*, Clarendon Press, Oxford. This is still the authoritative work on many aspects of lattice dynamics.

Slater, J. C. *Rev. mod Phys.* 1958, **30**, 197. This paper entitled 'Waves in Crystals' includes many historical references.

2

Lattices and Crystal Structure

2.1 Two-dimensional crystals

A lattice of points is a geometrical concept. In Fig. 2.1 we show the most general type of two-dimensional lattice, characterized by the lattice translations $\mathbf{a}_1$ and $\mathbf{a}_2$ which enclose an angle α_3. The area $a_1 a_2 \sin \alpha_3$ shown in the Figure

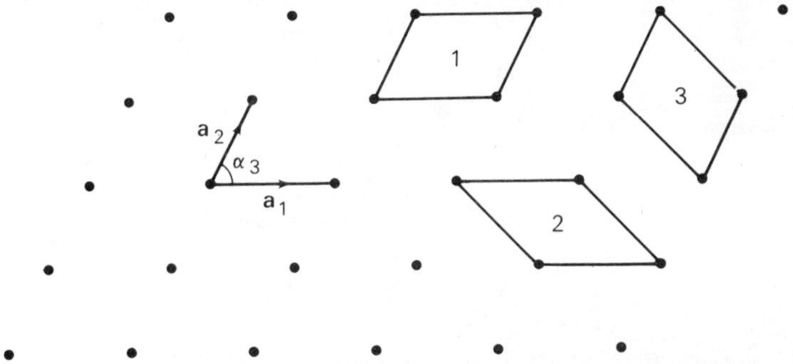

Figure 2.1 A small part of the most general type of two-dimensional lattice. The dimensions a_1, a_2 and α_3 define a unit cell denoted 1. Alternative choices, having the same area, are denoted 2 and 3.

is the area of the unit cell of the two-dimensional lattice. Although the parallelogram which is the unit cell can be chosen in different ways, a few of which are shown in Fig. 2.1, all have the same area and the one for which α_3 is closest to 90° is usually chosen. The two-dimensional analogue of the simplest possible type of crystal structure is now obtained by associating an atom with each unit cell, as shown in Fig. 2.2. Notice that in Fig. 2.2 there is no compelling reason why the origin of the unit cell, or lattice point, should be chosen to coincide with the single atom which is the content of one unit cell, although it may be convenient to take it to do so. Only a very peculiar and physically unrealistic force between atoms could lead to their crystallization in a plane with the structure shown in Fig. 2.2. A more likely arrangement is shown in Fig. 2.3. This array is much more symmetrical than the one shown in Fig. 2.2. The symmetry elements or operations present in an array of indefinite extent comprise fourfold and twofold rotation axes perpendicular

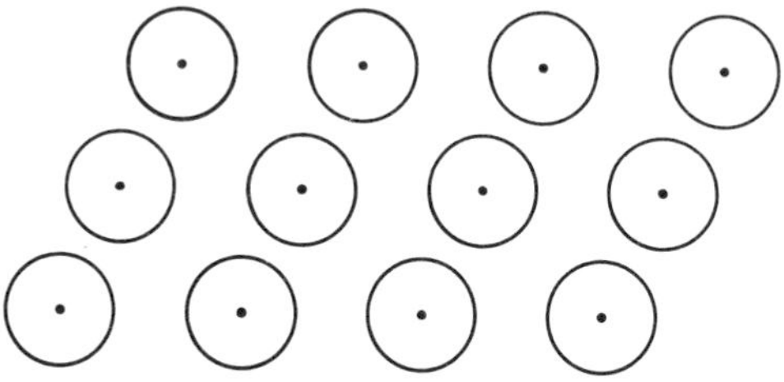

Figure 2.2 The simplest type of two-dimensional structure based on the lattice of Fig. 2.1.

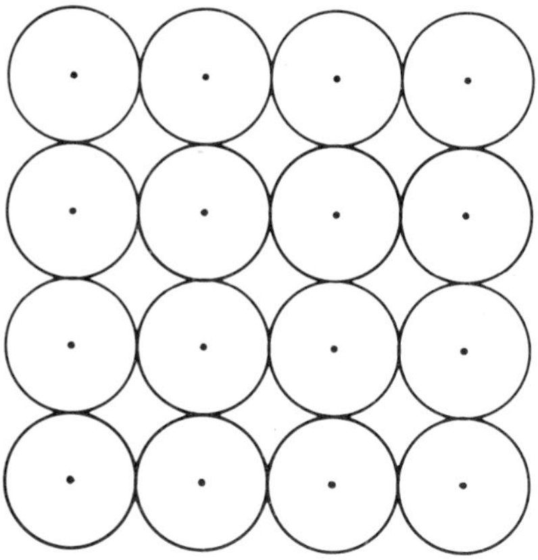

Figure 2.3 An array of atoms based on a square lattice, with one atom per unit cell.

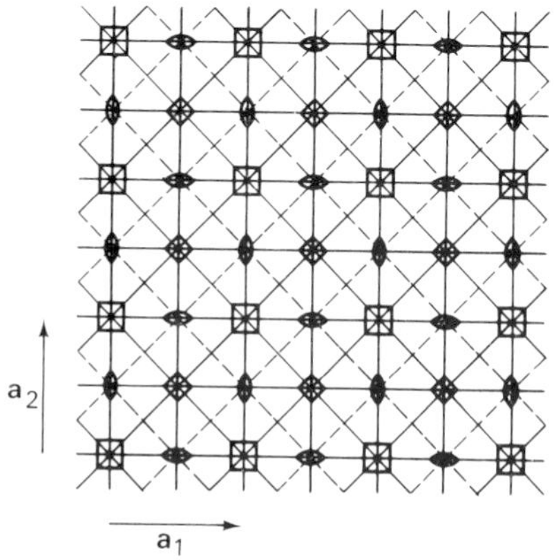

Figure 2.4 The symmetry elements of a square lattice. Lattice points coincide with four-fold axes denoted □.

to the plane, as well as mirror lines and glide lines in the plane. These are shown in Fig. 2.4; the operation of a glide line, shown dashed in the Figure, is a reflection followed by (in this case) a translation of $\frac{1}{2}(\mathbf{a}_1 + \mathbf{a}_2)$. The unit cell is also shown in Fig. 2.4. Notice that the symmetry elements can be imagined to exist quite independently of the atoms, that is, of the crystal structure, as is shown in Fig. 2.4. The unit cell is still obviously defined; a fourfold axis is a natural choice for the origin of the unit cell, or lattice point. This self-consistent arrangement of symmetry elements is called a plane group, or in three dimensions a space group. The number of different ways in which symmetry elements can be arranged in a plane to form a plane group is quite limited; there are only 17

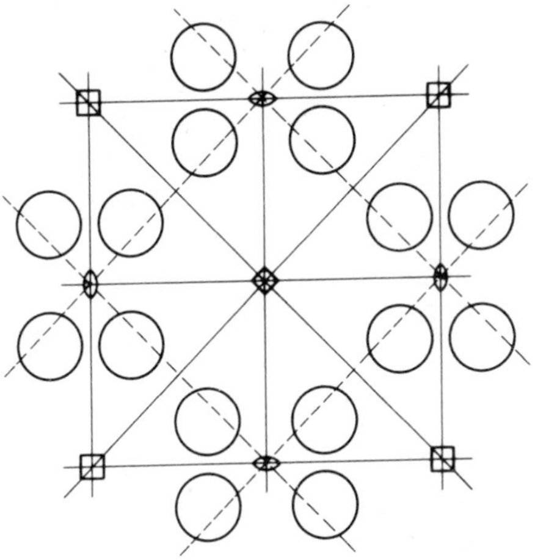

Figure 2.5 An atom placed in a general position of the unit cell of Fig. 2.4 is operated on by the symmetry elements so that there are necessarily eight atoms in one unit cell.

of them. They are illustrated in Buerger's *X-Ray Crystallography* (ref. 2.1) for example. In the 1890's, several years before the discovery of X-ray diffraction made it possible to assign a crystal structure to a particular space group, Fedorov, Shoenflies and Barlow showed independently that there are 230 space groups in three dimensions. These are described in Vol. I of *International Tables for Crystal Structure Determination* (ref. 2.2).

It it important to realize the distinction between the crystal structure and the lattice. The two-dimensional crystal structure shown in Fig. 2.3 has a one-to-one correspondence with the lattice, the atomic nuclei being the lattice points. This comes about because the atoms occupy *special positions* in the unit cell, in relation to the symmetry elements. For this to be possible each atom must have a certain minimum symmetry so that it can be reflected or rotated into itself. Suppose an atom is placed in a general position as shown in Fig. 2.5. It need not now possess symmetry. The atom is automatically multiplied up by the symmetry operations to give the structure shown, which has eight atoms in the unit cell. The set of points, in this case eight in number, is referred to as the set of general equivalent positions.

There are an infinity of crystal structures for each space group. The vibrations of the two-dimensional crystal shown in Fig. 2.5 would be quite different from those of the crystal shown in Fig. 2.3. The term 'lattice vibration' is therefore a somewhat unfortunate one, but is established by continual usage since 1912. Only in the simplest crystal structures of certain elements can the atomic nuclei be thought of as points of a single lattice. We could, of course, think of the structure shown in Fig. 2.5 as being made up of eight interpenetrating lattices, with the atoms at the lattice points of each, but this is not always a helpful concept.

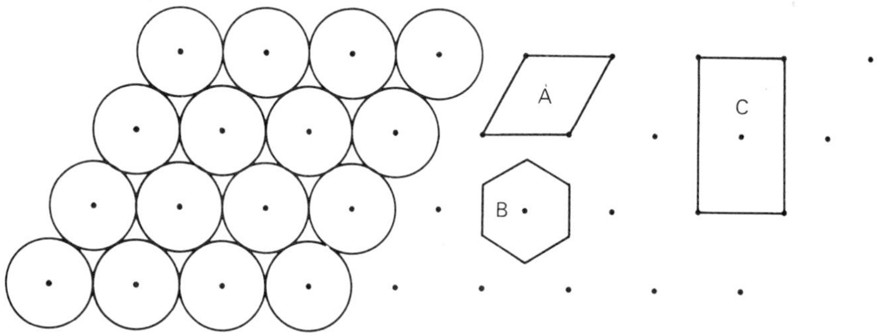

Figure 2.6 A close packed structure in two dimensions based on a hexagonal lattice. A, B and C are different ways of choosing the unit cell.

Fig. 2.6 illustrates the close packing of atoms in two dimensions. They are again at special positions, the sixfold rotation axes of a plane hexagonal lattice, and coincide with the lattice points. Three possible ways of choosing the unit cell are shown. A and B are called primitive unit cells since they contain only one lattice point. The non-primitive rectangular cell shown at C contains two lattice points. It is often convenient, as we shall see, to choose a non-primitive cell.

2.2 Cubic Lattices

The most general lattice in three dimensions is the triclinic lattice, characterized by three lattice translations $\mathbf{a}_1, \mathbf{a}_2, \mathbf{a}_3$ which are not mutually at right angles. The unit cell is therefore defined by the lengths a_1, a_2 and a_3 and the interaxial angles α_1, α_2 and α_3. In textbooks of crystallography the lattice is usually defined by vectors **a**, **b**, **c** and the interaxial angles are α, β, γ. We are not following this notation for reasons which will become clear later.

The only lattices which we shall discuss here in any detail are the three which are among the most important for solid state physics, at least at the present stage of development of the subject. These are the simple cubic, body-centred cubic and face-centred cubic lattices, the three Bravais lattices of the cubic system. The simple cubic lattice can be visualized without a diagram. The unit cell is primitive; although there is a lattice point at each of its eight corners, each point is shared by eight adjacent cells. If the lattice point is taken to be in the centre of the cell it is then obvious that the latter is primitive. There are 15 different space groups corresponding to the simple cubic lattice, each one having either 12, 24 or 48 general equivalent positions.

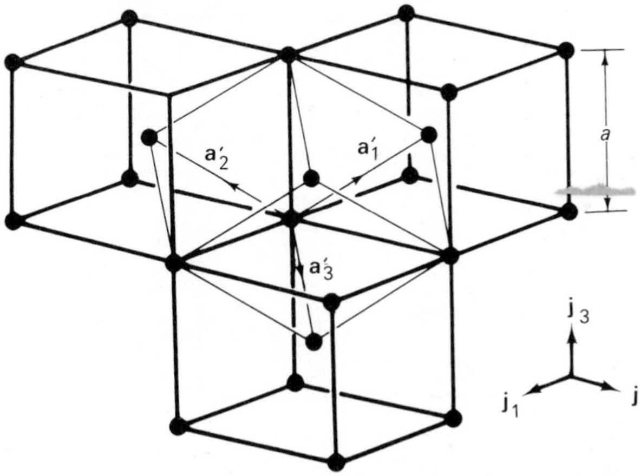

Figure 2.7 A body-centred cubic lattice can be referred to a trigonal unit cell with axes $\mathbf{a}'_1$, $\mathbf{a}'_2$, $\mathbf{a}'_3$.

Part of a body-centred cubic lattice is shown in Fig. 2.7, which shows two ways of choosing the unit cell. For the first, the unit cell is cubic, of side a, and contains two lattice points, at (000) and $(\frac{1}{2}\frac{1}{2}\frac{1}{2})$ in fractional coordinates. Let $\mathbf{j}_1\mathbf{j}_2\mathbf{j}_3$ be unit vectors parallel to the cube axes. Then as shown in Fig. 2.7 it is possible to choose new axes,

$$\begin{aligned} \mathbf{a}'_1 &= \tfrac{1}{2}a(-\mathbf{j}_1 + \mathbf{j}_2 + \mathbf{j}_3), \\ \mathbf{a}'_2 &= \tfrac{1}{2}a(\mathbf{j}_1 - \mathbf{j}_2 + \mathbf{j}_3), \\ \mathbf{a}'_3 &= \tfrac{1}{2}a(\mathbf{j}_1 + \mathbf{j}_2 - \mathbf{j}_3) \end{aligned} \tag{2.1}$$

which show that the body-centred cubic lattice is identical with a primitive trigonal lattice for which the interaxial angle is 109°28′. While this gives a primitive unit cell, it does not bring out the essentially cubic symmetry of the lattice and for most purposes one chooses the non-primitive cubic unit cell. There are 10 different space groups possible with this lattice; clearly when referred to the cubic unit cell the points $(x_1 x_2 x_3)$ and $(\frac{1}{2} + x_1, \frac{1}{2} + x_2, \frac{1}{2} + x_3)$ must be equivalent positions. Still another primitive unit cell, somewhat analogous to B of Fig. 2.6, can be found in the following way. Lines are drawn from a lattice point to its eight nearest neighbours, and to its six next nearest neighbours. Planes which are the perpendicular bisectors of these lines then enclose the volume shown in Fig. 2.8, a truncated octahedron which is called the Wigner-Seitz cell for this lattice. The Wigner–Seitz cell is a primitive unit cell with the origin at the centre of the cell.

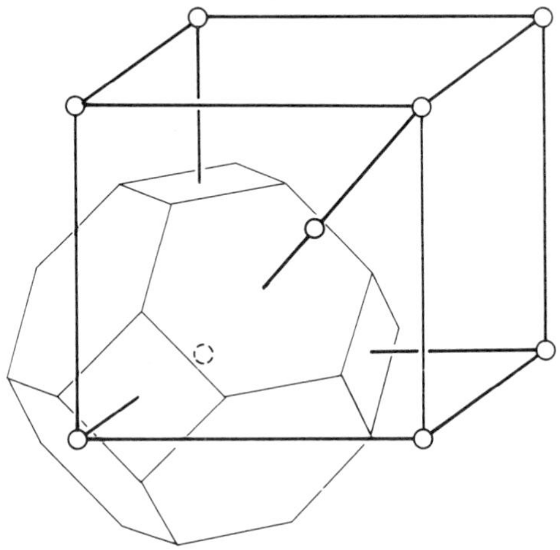

Figure 2.8 The relation of the Wigner–Seitz cell to the body-centred cubic unit cell.

A section of a face-centred cubic lattice is shown in Fig. 2.9, together with the primitive trigonal cell. In this case the trigonal axes are mutually inclined at 60°. Using a notation similar to that in Equ. 2.1, they are given by

$$\mathbf{a}_1'' = \tfrac{1}{2}a(\mathbf{j}_2 + \mathbf{j}_3), \quad \mathbf{a}_2'' = \tfrac{1}{2}a(\mathbf{j}_1 + \mathbf{j}_3), \quad \mathbf{a}_3'' = \tfrac{1}{2}a(\mathbf{j}_1 + \mathbf{j}_2) \tag{2.2}$$

There are 11 different space groups corresponding to the face-centred cubic lattice. The Wigner-Seitz cell of the face-centred cubic lattice is obtained by drawing lines to the twelve nearest neighbours of a lattice point. The planes which are the perpendicular bisectors of these lines enclose a rhombic dodecahedron which is also a primitive unit cell. (In this instance planes bisecting lines to next nearest neighbours all fall outside the closed volume just defined). The relation of the Wigner-Seitz cell to the face-centred cell is shown in Fig. 2.10.

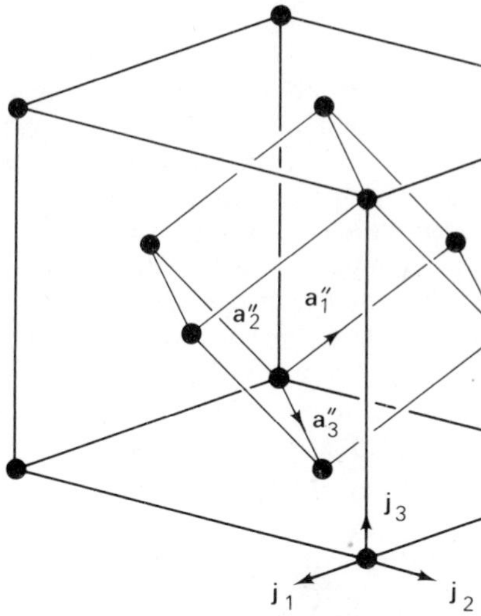

Figure 2.9 A face-centred cubic lattice can be referred to a trigonal unit cell with axes a_1'', a_2'', a_3''.

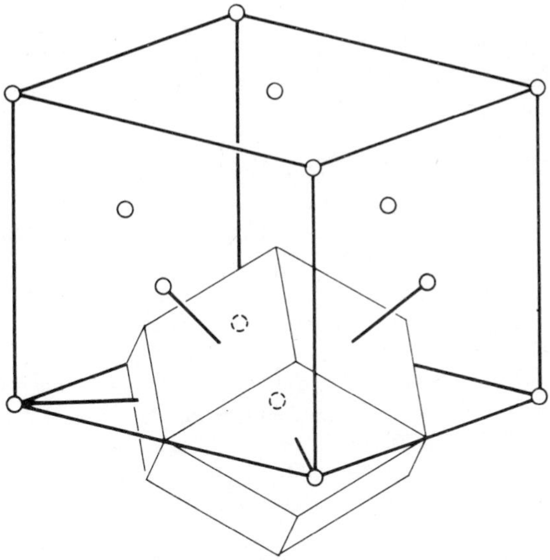

Figure 2.10 The relation of the Wigner–Seitz cell to the face-centred cubic unit cell.

2.3 Some simple crystal structures

Let us begin with a structure which does not exist in nature, we might call it the crystal structure of the element cubium. It is obtained by placing an atom at the origin of the simple cubic unit cell. Evidently this structure does not exist because it is unstable against a shearing deformation. The simplest structure based on the simple cubic lattice is the CsCl type of structure shown in Fig. 2.11. The atoms are in special positions in the unit cell, (000) for Cs and $(\frac{1}{2}\frac{1}{2}\frac{1}{2})$ for Cl. It is important to realize that this is *not* a structure based on the body-centred cubic lattice, since the points (000) and $(\frac{1}{2}\frac{1}{2}\frac{1}{2})$ are not equivalent, it is a structure based on the simple cubic lattice with two atoms per unit cell, in highly special

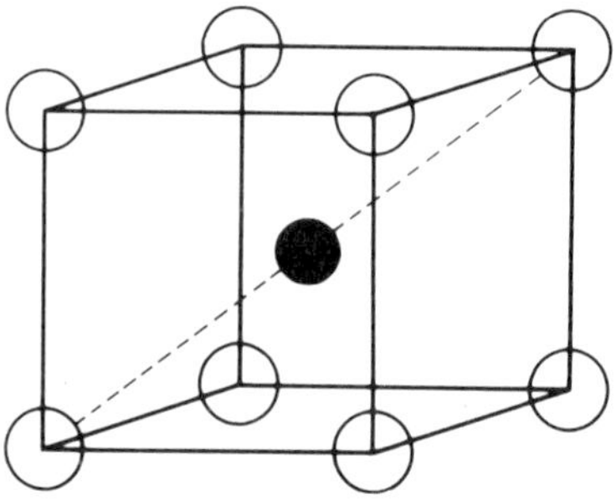

Figure 2.11 The structure of CsCl.

positions. Another structure for which the lattice is simple cubic is the so-called perovskite structure, of which $BaTiO_3$ (above 130°C) is a typical member. The atomic coordinates are

Ba at (000), Ti at $(\frac{1}{2}\frac{1}{2}\frac{1}{2})$, O at $(\frac{1}{2}\frac{1}{2}0)$, $(0\frac{1}{2}\frac{1}{2})$, $(\frac{1}{2}0\frac{1}{2})$,

and there are five atoms in the unit cell. The space group is the same for all these structures, namely Pm3m.

The crystal structure of sodium and of the other alkali metals is usually referred to as *the* body-centred cubic structure. The atoms are placed on the lattice points of the body-centred cubic lattice; there are therefore two atoms in the cubic unit cell, and one in the trigonal unit cell or in the Wigner–Seitz cell. The space group is Im3m. Just to remind ourselves that there are many crystal structures based on this lattice, Fig. 2.12 is a drawing of the crystal structure of hexamethylene tetramine $C_6H_{12}N_4$. The atoms are all in special positions, but special positions of lower symmetry than we have considered before. The space group is $I\bar{4}3m$.

Copper and certain other metals have *the* face-centred cubic structure, obtained by placing an atom at each point of the face-centred cubic lattice.

There are therefore four atoms in the cubic cell, at the equivalent points (000), $(\frac{1}{2}\frac{1}{2}0)$, $(\frac{1}{2}0\frac{1}{2})$, $(0\frac{1}{2}\frac{1}{2})$. The primitive trigonal cell and the Wigner–Seitz cell each have an atom at the origin. The inert elements such as argon and krypton also have this crystal structure, which is one of two different ways of close packing equal spheres. Each layer of atoms perpendicular to a body diagonal of the cubic cell exhibits the two-dimensional close packing shown in Fig. 2.6, but a

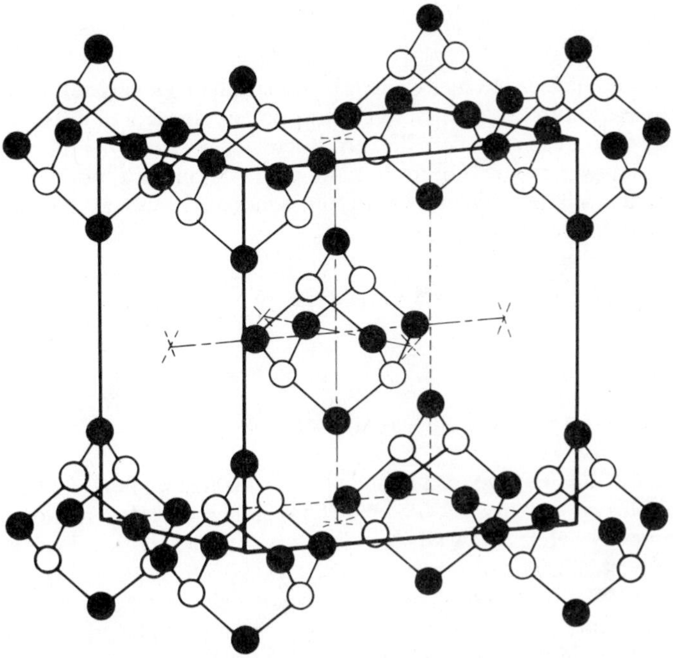

Figure 2.12 The crystal structure of hexamethylene tetramine $C_6H_{12}N_4$ is based on a body-centred cubic lattice. Each carbon (black in diagram) is bonded to two hydrogens, which are not shown.

three-dimensional model is really required to appreciate the fact that the atoms are close packed in three dimensions.

Atoms are, incidentally, less closely packed in the body-centred cubic structure. The structure of sodium chloride and of most other alkali halides is based on the face-centred cubic lattice with Na at (000), $(\frac{1}{2}\frac{1}{2}0)$, $(\frac{1}{2}0\frac{1}{2})$, $(0\frac{1}{2}\frac{1}{2})$, and Cl at $(\frac{1}{2}00)$, $(0\frac{1}{2}0)$, $(00\frac{1}{2})$, $(\frac{1}{2}\frac{1}{2}\frac{1}{2})$. Referred to the primitive trigonal unit cell, there are two atoms per cell, Na at (000) and Cl at $(\frac{1}{2}\frac{1}{2}\frac{1}{2})$. Both copper, and sodium chloride, belong to the space group Fm3m.

The diamond structure, which is also the structure of germanium and of silicon, has atoms at the points (000) $(\frac{1}{4}\frac{1}{4}\frac{1}{4})$ and the six other equivalent points of a structure based on the face-centred cubic lattice. The space group is Fd3m and the two atoms at (000), $(\frac{1}{4}\frac{1}{4}\frac{1}{4})$ are related by a centre of symmetry. In the trigonal unit cell there are two atoms at (000) and $(\frac{1}{4}\frac{1}{4}\frac{1}{4})$, or if the origin is moved to the centre of symmetry, they are at $\pm(\frac{1}{8}\frac{1}{8}\frac{1}{8})$. The structure of GaAs is somewhat similar, but since the two points are now occupied by inequivalent atoms the centre of symmetry is lost and the space group is $F\bar{4}3m$.

2.4 The reciprocal lattice

The reciprocal lattice has many applications in solid state physics, particularly in the theories of lattice dynamics, of electrons in crystals and of diffraction of radiation by crystals. (For a more detailed discussion see ref. 2.4.) It can be defined in a way which does not involve any particular physical problem. Let $\mathbf{a}_1, \mathbf{a}_2, \mathbf{a}_3$ define a lattice which we shall sometimes distinguish as the direct lattice. Now define

$$\mathbf{b}_1 = \frac{2\pi}{v}\,\mathbf{a}_2 \times \mathbf{a}_3, \quad \mathbf{b}_2 = \frac{2\pi}{v}\,\mathbf{a}_3 \times \mathbf{a}_1, \quad \text{and} \quad \mathbf{b}_3 = \frac{2\pi}{v}\,\mathbf{a}_1 \times \mathbf{a}_2 \tag{2.3}$$

where

$$v = \mathbf{a}_1 \cdot (\mathbf{a}_2 \times \mathbf{a}_3) = \mathbf{a}_2 \cdot (\mathbf{a}_3 \times \mathbf{a}_1) = \mathbf{a}_3 \cdot (\mathbf{a}_1 \times \mathbf{a}_2) \tag{2.4}$$

is the volume of the unit cell of the direct lattice. (Where we have used $\mathbf{b}_1, \mathbf{b}_2, \mathbf{b}_3$ crystallographers usually use $\mathbf{a}^*, \mathbf{b}^*, \mathbf{c}^*$.) The vectors $\mathbf{b}_1, \mathbf{b}_2, \mathbf{b}_3$ are not coplanar and consequently define a new lattice called the reciprocal lattice. Evidently when the interaxial angles of the direct lattice are 90°, $\mathbf{b}_1$ is parallel to $\mathbf{a}_1$ and $b_1 = 2\pi/a_1$ etc. The volume of the unit cell of the reciprocal lattice is easily seen in this instance to be $(2\pi)^3/v$, but this can be shown to be generally true. Another property of the translation vectors of the two lattices is

$$\begin{aligned} \mathbf{a}_i \cdot \mathbf{b}_i &= 2\pi \\ \mathbf{a}_i \cdot \mathbf{b}_j &= 0 \quad \text{for} \quad i \neq j \end{aligned} \tag{2.5}$$

The vector from the origin to a point $(l_1 l_2 l_3)$ of the direct lattice is

$$\mathbf{r}_l = l_1 \mathbf{a}_1 + l_2 \mathbf{a}_2 + l_3 \mathbf{a}_3 \tag{2.6}$$

where $\mathbf{r}_l$ is just a convenient shorthand notation, the subscript l implying the three integers $l_1 l_2 l_3$. Similarly in the reciprocal lattice we shall use $\mathbf{K}$ for a vector drawn from the origin, and

$$\mathbf{K}_h = h_1 \mathbf{b}_1 + h_2 \mathbf{b}_2 + h_3 \mathbf{b}_3 \tag{2.7}$$

is then the vector to the point $(h_1 h_2 h_3)$ of the reciprocal lattice. The usual crystallographic notation for $h_1 h_2 h_3$ is hkl but again we are not following it because we would soon run out of symbols.

From Equ. 2.5 it follows that

$$\mathbf{K}_h \cdot \mathbf{r}_l = 2\pi(h_1 l_1 + h_2 l_2 + h_3 l_3) = 2\pi \times \text{integer} \tag{2.8}$$

Another relation between the two lattices which we shall quote without giving the proof is the following. The equation

$$\frac{h_1 x_1}{a_1} + \frac{h_2 x_2}{a_2} + \frac{h_3 x_3}{a_3} = m \tag{2.9}$$

where $h_1 h_2 h_3$ are fixed integers and m assumes all integral values, defines a family of planes in the direct lattice such that every lattice point lies on one such plane. The vector $\mathbf{K}_h$ to the point $(h_1 h_2 h_3)$ of the reciprocal lattice (Equ. 2.7) is perpendicular to this family of planes and

$$K_h = \frac{2\pi}{d_h} \tag{2.10}$$

where d_h is the perpendicular distance between successive planes. The planes defined by Equ. 2.9 are always referred to as the $(h_1 h_2 h_3)$ planes (usually the (hkl) planes in crystallography). Thus for example the vector to the point (320) of the reciprocal lattice is perpendicular to the (320) planes, and its length is inversely proportional to the spacing of these planes, (see Fig. 2.13).

Obviously the reciprocal lattice of a simple cubic lattice with unit cell dimension a is also a simple cubic lattice in the same orientation and with unit cell dimension $2\pi/a$. What is less obvious is that the reciprocal lattice of a body-centred cubic lattice is a face-centred cubic lattice, and conversely. To see this, refer the b.c.c. lattice to trigonal axes, given by Equ. 2.1. Since this gives a primitive lattice, the corresponding reciprocal lattice is also primitive and is given by

$$\mathbf{b}_1' = \frac{2\pi}{v'} \mathbf{a}_2' \times \mathbf{a}_3' = \frac{\pi a^2}{v'} (\mathbf{j}_2 + \mathbf{j}_3) \tag{2.11}$$

using Equ. 2.1. But $v' = \frac{1}{2}a^3$, that is, the volume of the primitive cell is half that of the cubic cell. Hence

$$\mathbf{b}_1' = \frac{2\pi}{a} (\mathbf{j}_2 + \mathbf{j}_3) \tag{2.12}$$

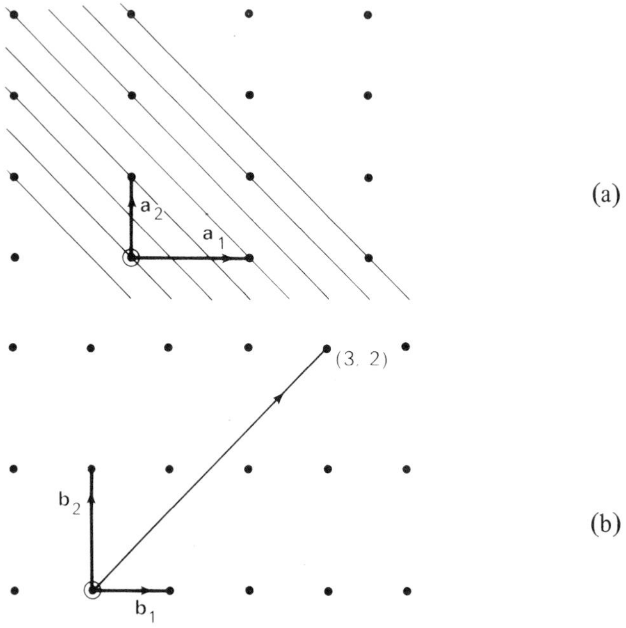

Figure 2.13 (a) The lines $(3x_1/a_1) + (2x_2/a_2) = m$ for $m = -1, 0, 1, \ldots, 6$, in a two-dimensional lattice. (b) The vector to the point (3, 2) of the reciprocal lattice is perpendicular to the corresponding family of lines shown above. Its length is inversely proportional to their spacing.

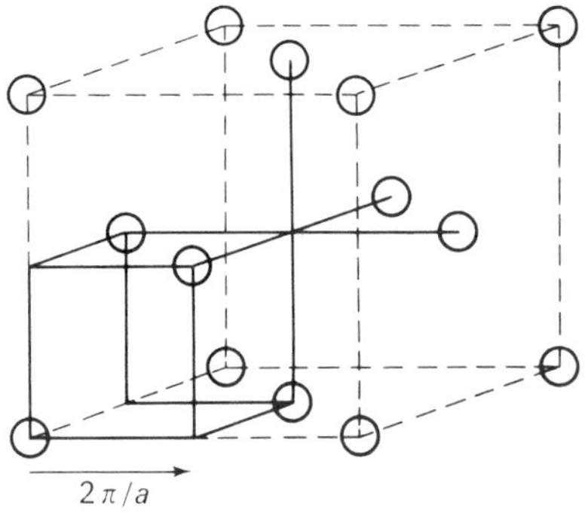

Figure 2.14 The points of a face-centred cubic lattice of side $4\pi/a$ can be referred to a simple cubic unit cell of side $2\pi/a$, but with half the points of the simple cubic lattice missing.

and similarly

$$\mathbf{b}_2' = \frac{2\pi}{a}(\mathbf{j}_1 + \mathbf{j}_3), \quad \mathbf{b}_3' = \frac{2\pi}{a}(\mathbf{j}_1 + \mathbf{j}_2) \tag{2.13}$$

Comparison with Equ. 2.2 shows that $\mathbf{b}_1'$, $\mathbf{b}_2'$, $\mathbf{b}_3'$ are the trigonal axes of a f.c.c. lattice whose unit cell dimension is $4\pi/a$. Since we began with a direct lattice having cell dimension a we might have expected the reciprocal lattice to have a cell dimension $2\pi/a$ rather than $4\pi/a$. We can of course describe a f.c.c. lattice of cell dimension $4\pi/a$ as a simple cubic lattice of cell dimension $2\pi/a$, provided we say that points $(h_1 h_2 h_3)$ for which $h_1 + h_2 + h_3$ is odd are 'systematically absent' and this is what is always done in crystallography. The situation is illustrated in Fig. 2.14.

Similarly we find that, starting from a f.c.c. direct lattice of cell dimension a we obtain a b.c.c. reciprocal lattice of cell dimension $4\pi/a$ and if we refer the latter to a simple cubic lattice of side $2\pi/a$ we must retain only those points $(h_1 h_2 h_3)$ for which the integers are either all odd or all even.

The unit cell in the reciprocal lattice is almost invariably chosen as the Wigner–Seitz cell, with the point (000) at the centre of the cell. While the name 'Wigner–Seitz cell' is used in the direct lattice, the same entity in the reciprocal lattice is called the 'first Brillouin zone', or often simply the Brillouin zone. Thus a b.c.c. direct lattice, which has a f.c.c. reciprocal lattice, has the Wigner–Seitz cell shown in Fig. 2.8 for the direct lattice and the Brillouin zone shown in Fig. 2.10 for its reciprocal lattice.

Notice that in this discussion we have not found it necessary to mention the properties of crystals, only the properties of lattices.

2.5 X-ray diffraction

It is not necessary for our present purpose to discuss the physical processes involved in the scattering of X-rays. We shall simply take the amplitude of the radiation scattered coherently in a particular direction by a single electron as our unit. The wavelength of the scattered radiation in this process is the same as that of the incident radiation. Let $\mathbf{k}_0$ be a vector of magnitude $2\pi \div$ wavelength in the direction of the incident X-ray beam, and correspondingly let $\mathbf{k}_1$ be the wavevector of the scattered radiation. Since there is no change of wavelength, $k_1 = k_0$. The scattering vector is defined as

$$\mathbf{K} = \mathbf{k}_1 - \mathbf{k}_0 \tag{2.14}$$

and if the angle of scattering is 2θ we see from Fig. 2.15 that $K = 4\pi/\lambda\,(\sin\theta)$.

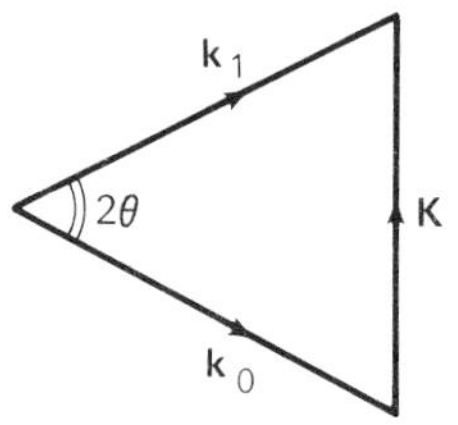

Figure 2.15 Since $k_0 = k_1 = (2\pi/\lambda)$, $K = (4\pi/\lambda)(\sin\theta)$, **K** is the scattering vector.

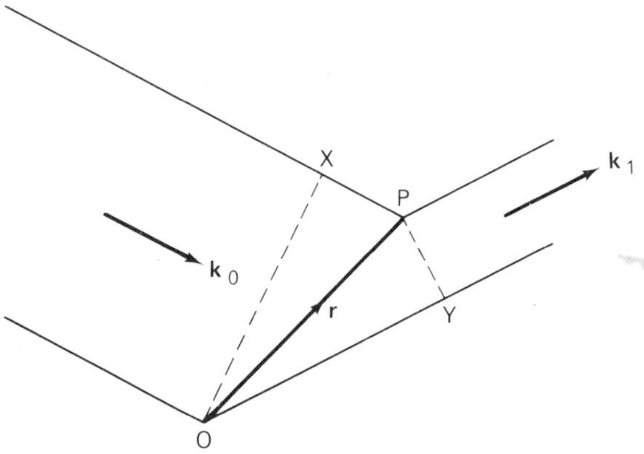

Figure 2.16 This diagram illustrates the path difference OY – PX between radiation scattered from the origin O and from a point P at **r**.

From Fig. 2.16 we see that if an electron at the origin scatters a wave of unit amplitude, an electron at **r** scatters with a phase difference

$$= \frac{2\pi}{\lambda}(\mathrm{OY} - \mathrm{PX}) = \mathbf{k}_1 \cdot \mathbf{r} - \mathbf{k}_0 \cdot \mathbf{r} = \mathbf{K} \cdot \mathbf{r} \tag{2.15}$$

and its contribution to the amplitude of the scattered beam will be $\exp(i\mathbf{K} \cdot \mathbf{r})$. Consequently the amplitude of the beam scattered in the direction of $\mathbf{k}_1$ by a number n of electrons is

$$A(\mathbf{K}) = \sum_{j=1}^{n} \exp(i\mathbf{K} \cdot \mathbf{r}_j) \tag{2.16}$$

and for a continuous distribution of electron density $\rho(\mathbf{r})$,

$$A(\mathbf{K}) = \int \rho(\mathbf{r}) \exp(i\mathbf{K} \cdot \mathbf{r}) d^3r \tag{2.17}$$

(d^3r denotes an element of volume at the point $\mathbf{r}$). In words, the (complex) amplitude of the scattered radiation is the Fourier transform of the electron distribution.

Suppose the distribution $\rho(\mathbf{r})$ is that of a single atom, which we can take as a good approximation to be spherically symmetric. It can then be shown that Equ. 2.17 reduces to

$$A_a(\mathbf{K}) = \int 4\pi r^2 \rho(r) \frac{\sin Kr}{Kr} dr \tag{2.18}$$

This quantity is usually written $f(K)$, and is called the form factor or atomic scattering factor. It has no imaginary component and this means that for scattering through an angle such that the scattering vector is $\mathbf{K}$, we can mentally replace the entended distribution of the atom by $f(K)$ electrons at the nucleus. As K increases from zero, $f(K)$ decreases from the atomic number, to zero for short wavelength radiation scattered through a large angle.

Next we consider scattering by a lattice, or since we have been emphasizing that the latter is a geometrical concept, let us give it scattering power by imagining an electron (our scattering unit) at each lattice point. We take the array to be a finite parallelepiped of dimensions $N_1\mathbf{a}_1, N_2\mathbf{a}_2, N_3\mathbf{a}_3$. Taking the origin as the centre of the array, the amplitude of the scattered radiation is

$$A_L(\mathbf{K}) = \sum \exp i\mathbf{K} \cdot (l_1\mathbf{a}_1 + l_2\mathbf{a}_2 + l_3\mathbf{a}_3) \tag{2.19}$$

where the sum is over all values of l_1 between $-\frac{1}{2}(N_1 - 1)$ through zero to $+\frac{1}{2}(N_1 - 1)$, and similarly for l_2 and l_3. This can be evaluated as the product of three separate geometrical progressions, and the result is

$$A_L(\mathbf{K}) = \left(\frac{\sin \frac{1}{2}N_1\mathbf{K}\cdot\mathbf{a}_1}{\sin \frac{1}{2}\mathbf{K}\cdot\mathbf{a}_1}\right)\left(\frac{\sin \frac{1}{2}N_2\mathbf{K}\cdot\mathbf{a}_2}{\sin \frac{1}{2}\mathbf{K}\cdot\mathbf{a}_2}\right)\left(\frac{\sin \frac{1}{2}N_3\mathbf{K}\cdot\mathbf{a}_3}{\sin \frac{1}{2}\mathbf{K}\cdot\mathbf{a}_3}\right) \tag{2.20}$$

Now let us express the scattering vector $\mathbf{K}$ in terms of the translation vectors of the reciprocal lattice by writing

$$\mathbf{K} = \xi_1\mathbf{b}_1 + \xi_2\mathbf{b}_2 + \xi_3\mathbf{b}_3 \tag{2.21}$$

This is just a matter of convenience, we can refer a vector to any set of axes we care to choose. When $\xi_1\xi_2\xi_3$ are all integers the vector $\mathbf{K}$ will end on a point of the reciprocal lattice, but we can always think of $\mathbf{K}$ as a vector in the space of the reciprocal lattice, in reciprocal space for short. By inspection of Equ. 2.20 we see that when $\xi_1\xi_2\xi_3 = h_1h_2h_3$, i.e. when $\mathbf{K} = \mathbf{K}_h$ (See Equ. 2.7),

$$A_L(\mathbf{K}_h) = \left(\frac{\sin N_1\pi h_1}{\sin \pi h_1}\right)\left(\frac{\sin N_2\pi h_2}{\sin \pi h_2}\right)\left(\frac{\sin N_3\pi h_3}{\sin \pi h_3}\right) = N_1N_2N_3 \tag{2.22}$$

The amplitude is thus equal to the total number of electrons present; they must therefore all be scattering in phase with one another when the condition $\mathbf{K} = \mathbf{K}_h$ is satisfied. Using the relation mentioned earlier, between planes in the direct lattice and points of the reciprocal lattice, it is possible to show that the condition we have established is equivalent to Bragg's law, where the scattered beam is thought of as having been produced by reflection of the incident beam from a set of planes.

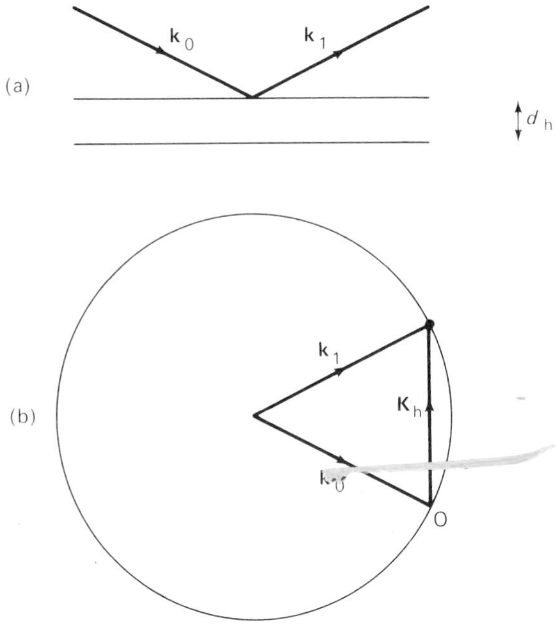

Figure 2.17 Two aspects of X-ray diffraction. (a) The beam is reflected from a set of planes of spacing d_h. (b) In terms of the reciprocal lattice, the vector $\mathbf{K}_h$, which is perpendicular to the set of planes and is of length $(2\pi/d_h)$, must end on the surface of the 'sphere of reflection'.

The condition $\mathbf{K} = \mathbf{K}_h$ for all the lattice points to scatter in phase can be expressed geometrically as shown in Fig. 2.17. The vector $\mathbf{k}_0$ will usually be fixed in magnitude and direction, but $\mathbf{k}_1$ is fixed only in magnitude and can terminate anywhere on a sphere of radius $\mathbf{k}_0 = 2\pi/\lambda$ whose centre is the origin for $\mathbf{k}_0$ and $\mathbf{k}_1$. The scattering vector $\mathbf{K}$ can therefore be drawn from the point O (Fig. 2.17) to any other point on the surface of the sphere. Now make O the origin of the reciprocal lattice. To satisfy the condition $\mathbf{K} = \mathbf{K}_h$ a point of the reciprocal lattice must fall on the surface of the sphere. Thus in general when a beam of monochromatic X-rays falls on our array of electrons there will be no scattered beam. When, however, the array is turned round, its reciprocal lattice, from the way it was defined, turns with it, and a reciprocal lattice point will eventually touch the sphere and a scattered beam will appear in the direction of the vector $\mathbf{k}_1$. Every point for which $K_h < 4\pi/\lambda$ can obviously be brought on to the sphere by appropriate choice of orientation. Using this construction, it is easier to visualize X-ray diffraction in terms of the reciprocal lattice than in terms of reflection by planes of the direct lattice.

Next we consider the question, how does the amplitude of the diffracted beam decrease as $\mathbf{K}$ deviates from $\mathbf{K}_h$. Evidently when $N_1N_2N_3$ are large the decrease is very sudden. We see from Equ. 2.20 that $\mathbf{K}$ has only to change by $2\pi/N_1a_1$ in the $\mathbf{a}_1$ direction for the amplitude $A_L(\mathbf{K})$ to fall to zero, and it remains essentially zero until the next reciprocal lattice point is reached. It is not difficult to show that the 'scattering power' in reciprocal space is confined to a volume $(2\pi)^3/V$ around each reciprocal lattice point, where $V = N_1N_2N_3v$ is the volume of the scattering array of electrons. Thus as $N_1N_2N_3$ increases the function $A_L(\mathbf{K})$ becomes higher at each reciprocal lattice point and more narrowly confined around it. It is readily shown from Equ. 2.20 that its value integrated over a volume in the vicinity of a reciprocal lattice point remains constant, however,

$$\int_{\mathbf{K}\text{ near }\mathbf{K}_h} A_L(\mathbf{K})d^3K = \frac{(2\pi)^3}{v} \tag{2.23}$$

Thus, if we introduce a three-dimensional Dirac δ-function having the properties

$$\delta(\mathbf{K}) = 0 \quad \text{except when} \quad \mathbf{K} = 0$$

and

$$\int_{\mathbf{K}\text{ near }0} \delta(\mathbf{K})d^3K = 1 \tag{2.24}$$

in the limit $N_1N_2N_3 \to \infty$ we can rewrite Equ. 2.20 as

$$A_L(\mathbf{K}) = \frac{(2\pi)^3}{v} \sum_h \delta(\mathbf{K} - \mathbf{K}_h) \tag{2.25}$$

This is just the mathematical expression of the fact that the amplitude of the X-ray beam scattered by a lattice which has an electron at every lattice point, is zero unless the scattering vector $\mathbf{K}$ is also a vector to a point of the reciprocal lattice. Another way of expressing this result, which makes no mention of X-ray diffraction, is that the Fourier transform (as defined by Equ. 2.17) of an infinite lattice of points of unit weight is a reciprocal lattice of points of weight $(2\pi)^3/v$.

We now bring the discussion somewhat closer to reality by considering X-ray diffraction by a crystal rather than by a single electron at each lattice point. To convert the lattice to a crystal we must put the same continuous electron distribution in each unit cell. The distribution in each unit cell will have the symmetry imposed by the space group, but this need not concern us at present. We simply take the electron density in any one unit cell referred to the origin of that cell to be $\sigma(\mathbf{r})$. The electron density in the crystal is then given by

$$\rho_c(\mathbf{r}) = \sum_l \sigma(\mathbf{r} - \mathbf{r}_l) \tag{2.26}$$

i.e., it is the sum of that in the separate cells. We are using the notation introduced at Equ. 2.6. The dimensions of the crystal are the same as before, that is $-\frac{1}{2}(N_1 - 1) \leqslant l_1 \leqslant \frac{1}{2}(N_1 - 1)$, etc. From Equ. 2.17, the amplitude scattered by the crystal is therefore

$$A_c(\mathbf{K}) = \int \rho_c(\mathbf{r}) \exp(i\mathbf{K}\,.\,\mathbf{r}) d^3 r \tag{2.27}$$

the integral being over the volume of the crystal. Using Equ. 2.26 it is found after some algebra (which we shall not give) that Equ. 2.27 leads to

$$A_c(\mathbf{K}) = \frac{(2\pi)^3}{v} \sum_h F(\mathbf{K})\, \delta(\mathbf{K} - \mathbf{K}_h) \tag{2.28}$$

where the *structure factor* $F(\mathbf{K})$ is defined by

$$F(\mathbf{K}) = \int_{\text{cell}} \sigma(\mathbf{r}) \exp(i\mathbf{K}\,.\,\mathbf{r}) d^3 r \tag{2.29}$$

The geometrical conditions for a diffracted beam to be produced are therefore just the same as we found before, namely that the scattering vector $\mathbf{K}$ must coincide with a vector of the reciprocal lattice $\mathbf{K}_h$. The new feature is that the amplitude of the diffracted beam produced when $\mathbf{K} = \mathbf{K}_h$ is proportional to the structure factor $F(\mathbf{K}_h)$, and thus depends on the electron distribution in any one unit cell. Note that although the structure factor can be defined for a general value of $\mathbf{K}$ (Equ. 2.29), the form of Equ. 2.28 ensures that it can be 'observed' only at reciprocal lattice points. Remembering that $\mathbf{K}_h$ is specified by three integers h_1, h_2 and h_3 for which h is a shorthand notation, $F(\mathbf{K}_h)$ could equally well be written F_h, or $F(h_1 h_2 h_3)$. In practice we can usually simplify Equ. 2.29 by taking the electron density $\sigma(\mathbf{r})$ to be a superposition of that of the individual spherically symmetric atoms in one unit cell.

Denoting the atomic scattering factor of an atom of type κ by $f_\kappa(K)$ (Equ. 2.18) it is found that Equ. 2.29 then reduces to

$$F(\mathbf{K_h}) = \sum_{\kappa=1}^{n} f_\kappa(K) \exp(i\mathbf{K_h} \cdot \mathbf{r}_\kappa) \qquad (2.30)$$

where $\mathbf{r}_\kappa$ is the position of an atom in the unit cell. We shall find in Chapter 5 that one outcome of thermal vibration is in effect to modify the atomic scattering factor, replacing $f_\kappa(K)$ by $f_\kappa(K) \exp(-\frac{1}{2}K^2 \overline{u_\kappa^2})$ where $\overline{u_\kappa^2}$ is the mean square displacement of the atom which results from thermal vibration. We shall also see in Chapter 5 that very similar formulae apply when we are considering neutron diffraction rather than X-ray diffraction.

In practice the amplitude and phase of a diffracted beam cannot be directly measured, only the intensity, so that we have

$$|F(\mathbf{K_h})|^2 = \left| \sum_{\kappa=1}^{n} f_\kappa(K) \exp(i\mathbf{K_h} \cdot \mathbf{r}_\kappa) \right|^2 \qquad (2.31)$$

The problem of determining the atomic coordinates $\mathbf{r}_\kappa$ from the measured intensities $|F(\mathbf{K_h})|^2$ is clearly not a simple one when the crystal structure is at all complicated, but the topic is not one we wish to pursue here and the reader is referred to textbooks on crystal structure determination, (refs 2.3, 2.4).

The derivations in §2.5 are somewhat more algebraic than any we have met so far, so it is desirable to summarize the essential features. The most important is that a lattice, and a perfect crystal, scatter X-rays only in narrowly defined directions given by

$$\mathbf{K} = \mathbf{k}_1 - \mathbf{k}_0 = \mathbf{K_h}$$

That is, the difference in wavevectors of the scattered and incident beams must be a vector of the reciprocal lattice. This geometrical condition applies whatever the distribution of electrons in the unit cell, provided it is the same in every unit cell. However the *intensity* of each diffracted beam is dependent on the electron distribution, that is, on the crystal structure.

References and suggestions for further reading

2.1 Buerger, M. J. *X-ray Crystallography*, Wiley and Sons, 1942.
2.2 *International Tables for Crystal Structure Determination*, Vol. I, 1952, Ed. K. Lonsdale, Kynoch Press, Birmingham.
2.3 Lipson, H. and Cochran, W. 1966, *The Determination of Crystal Structures*, G. Bell and Sons, London.
2.4 The subject matter of this chapter and a more extended discussion of symmetry of real crystals and their study by diffraction processes will be found in Brown, P. J. and Forsyth, J. B. 1973, *The Crystal Structure of Solids*, Edward Arnold, London, a companion volume in the series to which the present book belongs.

3

Dynamics of One-dimensional Crystals

3.1 The dispersion relation

When certain approximations are made, the dynamics of a three-dimensional crystal is not a particularly difficult subject, but it is a very algebraic one with a profusion of indices and suffices and it is easy to lose one's way among them. Many of the essential features can be brought out by considering a one-dimensional crystal or linear chain.

The simplest type of one-dimensional vibrating system with many degrees of freedom is a stretched wire with fixed ends. The normal modes of vibration of this system (constrained to move in a plane) are standing waves with an integral number of half waves fitted into the length of the wire. The amplitude of each such standing wave is a normal coordinate of the system and the total energy of the vibrating system is the sum of the energies of the different modes of vibrations simultaneously present. The frequency of each mode of vibration is inversely proportional to the corresponding wavelength, i.e. proportional to wavenumber. Further discussion of this system will be found in textbooks on wave motion, for example Coulson (ref. 3.1) but it is assumed that the reader only requires to be reminded of its properties. Boundary conditions which are somewhat more convenient mathematically, if less realistic physically, can be obtained by imagining the wire joined into a continuous circle–we ignore the difficulty of keeping it under tension while leaving it free to vibrate! The normal modes of vibration are then waves travelling in either direction round the circle, each with a whole number of wavelengths fitted into the circumference. We shall use this type of boundary condition in our discussion of the linear chain or one-dimensional crystal. We begin with the simplest type of crystal, one which contains only one atom of mass m in a unit cell of dimension a.

We assume that forces between the atoms extend only to nearest neighbours, and that the crystal is 'harmonic' in the sense mentioned in Chapter 1, so we represent the interaction between nearest neighbours by a spring of force constant, f, Fig. 3.1. We take the vibrations to be longitudinal, that is the displacement of an atom is along the length of the chain. Each atom is identified by an index l so that its equilibrium position measured from some convenient origin is $r_l = la$, and l can assume the values, $0, 1, \ldots, N-1$.

Taking the displacement of the l'th atom to be u_l, its equation of motion is

$$m\frac{\partial^2 u_l}{\partial t^2} = f\{u_{l+1} - u_l + u_{l-1} - u_l\} \tag{3.1}$$

Guided by the result which applies for a vibrating wire, we assume that there is a solution of the form

$$u_l = (Nm)^{-1/2}|B(q)| \cos(qr_l - \omega(q)t + \alpha(q)) \tag{3.2}$$

The factor $(Nm)^{-1/2}$ is convenient for normalizing purposes. We shall sometimes ignore its presence in referring to $|B(q)|$ as the amplitude of the wave.

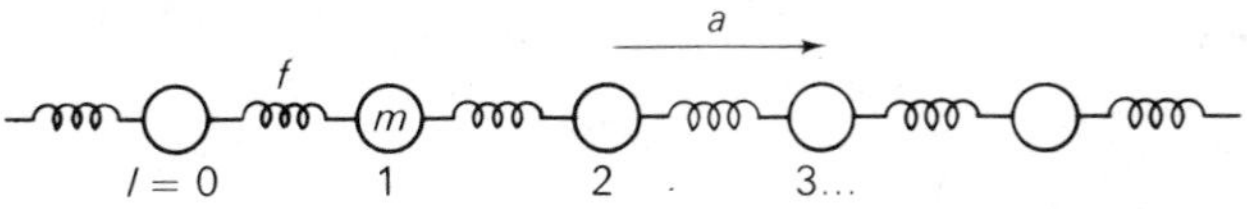

Figure 3.1 A one-dimensional crystal, or linear chain.

Correspondingly, q, $\omega(q)$ and $\alpha(q)$ are obviously the wavenumber, (angular) frequency and initial phase of a wave travelling to the right. $|B(q)|$ and $\alpha(q)$ will be determined by the initial conditions. We write $|B(q)|$ to emphasize that it is a real number. It is often convenient† to introduce a complex amplitude

$$B(q) = |B(q)| \exp(i\alpha(q)) \tag{3.3}$$

On substituting Equ. 3.2 in Equ. 3.1, one finds that the equation is satisfied provided that

$$m\omega^2(q) = 2f(1 - \cos qa) = 4f \sin^2 \tfrac{1}{2}qa \tag{3.4}$$

This relation between ω and q is called the dispersion relation or dispersion curve. For modes of vibration of long wavelength, i.e. small value of q, the result reduces to

$$\omega(q) = qa(f/m)^{1/2} \tag{3.5}$$

which is the same result as for a wire with tension fa and linear density m/a. Since the wavelength is large compared with the lattice spacing in this limit, a disturbance of this sort must clearly propagate like a longitudinal compressional wave in a continuum and the velocity defined by the initial slope $d\omega/dq$ is thus the velocity of sound. The dispersion relation is not linear for larger values of q.

†See p. 33 for example.

The frequency rises to a maximum of $2(f/m)^{1/2}$ at $q = \pi/a$ and falls to zero at $q = 2\pi/a$, in fact $\omega(q)$ is periodic with periodicity $2\pi/a$, which we recognize as the dimension of the one-dimensional reciprocal lattice, (Fig. 3.2).

In Fig. 3.2 we have shown the dispersion relation as if all values of q were possible. However, it is apparent from Equ. 3.2, remembering that $r_l = la$, that increasing q by an integral number times $2\pi/a$ makes absolutely no difference to the displacement u_l of an atom, nor does it give a wave of different frequency as we see from Fig. 3.2. In this respect waves on a linear chain have quite different properties from waves in a continuum, such as a wire, and we obtain all possible patterns of displacement and all possible frequencies by restricting q to the range $\pm\,\pi/a$, i.e. q can be taken to be in the (first) Brillouin zone. Notice that $+q$ and $-q$ are not equivalent since one corresponds to a wave travelling to

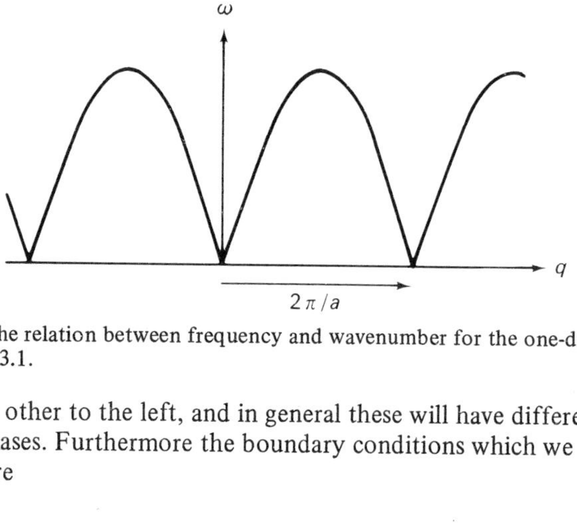

Figure 3.2 The relation between frequency and wavenumber for the one-dimensional crystal of Fig. 3.1.

the right, the other to the left, and in general these will have different amplitudes and initial phases. Furthermore the boundary conditions which we have chosen require

$$u_0 = u_N$$

which is only possible when an integral number s of wavelengths can be fitted into the length Na, i.e.

$$s\frac{2\pi}{q} = Na \quad \text{or} \quad q = s\frac{2\pi}{Na} \tag{3.6}$$

Thus successive allowed values of q are separated by $2\pi/Na$, and since $-\pi/a \leqslant q \leqslant +\pi/a$, it follows that exactly N different values of q are allowed. Thus the general solution for the motion of the atoms of the linear chain is

$$u_l = (Nm)^{-1/2} \sum_q |B(q)| \cos(qla - \omega(q)t + \alpha(q)) \tag{3.7}$$

the sum being over the range of values of q just indicated. There are $2N$ initial conditions, the initial position and velocity of every atom. This agrees with Equ. 3.7, since an amplitude $|B(q)|$ and phase $\alpha(q)$ will have to be determined for each of N independent values of q, by the initial conditions.

The kinetic energy of the system is given by

$$T = \tfrac{1}{2} m \sum_{l} \dot{u}^2(l) \tag{3.8}$$

Substituting from Equ. 3.7 and using the results that on averaging over the time t,

$$\sum_{l} |B(q)|^2 \cos^2(qla - \omega(q)t + \alpha(q)) = \tfrac{1}{2} N |B(q)|^2$$

while

$$\sum_{l} |B(q)|\,|B(q')| \cos(q'la - \omega(q')t + \alpha(q')) \cos(qla - \omega(q)t + \alpha(q)) = 0$$

one finds that

$$\overline{T} = \tfrac{1}{4} \sum_{q} \omega^2(q) |B(q)|^2 \tag{3.9}$$

A certain amount of algebra shows that the potential energy is given by the same expression so that the total energy of the system is

$$\overline{U} = \tfrac{1}{2} \sum_{q} \omega^2(q) |B(q)|^2 \tag{3.10}$$

What we have done so far is to make a transfer to a new coordinate system. Originally we were concerned with the specification of the displacements and velocities of each of the N atoms; now we have an equal number of independent normal modes, each one of which specifies the associated correlated displacements of all the atoms, and for each of which the amplitude and phase is determined by the initial conditions. Each independent wavenumber q specifies a normal mode of vibration of the system, and the energy is the sum of the energies of the different normal modes. It is often convenient to make a transformation to complex normal coordinates, in this instance by writing

$$u_l = (Nm)^{-1/2} \sum_{q} Q(q) \exp(iqr_l) \tag{3.11}$$

so that the time dependence is incorporated in the coordinate $Q(q)$. The range of q is as before, but evidently since u_l is a real displacement,

$$Q(-q) = Q^*(q) \tag{3.12}$$

A full discussion of complex normal coordinates is somewhat beyond the scope of this book, although it is important for a thorough treatment of the quantum mechanics of this system. Further discussion will be found in the review articles by Cochran and Cowley (ref. 3.2) and by Maradudin, Montroll and Weiss (ref. 3.3). Working out the energy of the system in terms of complex normal coordinates and getting the numerical factor correct is quite tricky. The result is

$$\bar{U} = \sum_q \omega^2(q)|Q(q)|^2$$

3.2 Quantization and phonons

For more rigorous discussions than we shall attempt here, the reader is referred to the articles mentioned at the end of the previous section, and to Jensen (ref. 3.4). In classical mechanics the theorem of equipartition of energy holds, and a system in equilibrium with its surroundings at temperature T has an energy $\frac{1}{2}k_BT$ per degree of freedom. Thus the average kinetic and potential energies of an oscillator are both $\frac{1}{2}k_BT$ and by analogy the average energy of one mode of vibration is

$$\bar{U}(q) = \tfrac{1}{2}\omega^2(q)|B(q)|^2 = k_BT \tag{3.13}$$

The total energy of a one-dimensional crystal is therefore

$$\bar{U} = \sum_q \bar{U}(q) = Nk_BT \tag{3.14}$$

and the specific heat $d\bar{U}/dT$ is therefore independent of temperature. The same conclusion follows from similar assumptions about a three-dimensional crystal, contrary to what is found in practice. As we remarked in Chapter 1 the way out of this difficulty was already apparent to Einstein in 1907.

There is a close analogy between the expression (3.10) and the energy of a simple harmonic oscillator. If we write the displacement of a mass m oscillating with frequency ω as

$$u = m^{-1/2} B \cos(\omega t + \alpha) \tag{3.15}$$

the energy is $\frac{1}{2}B^2\omega^2$. Thus in classical mechanics each normal mode of vibration behaves like an independent simple harmonic oscillator. Fortunately this is still true in quantum mechanics, at least for a harmonic crystal, and we need only discuss briefly the quantization of a simple harmonic oscillator. It is a well known result (see for example Schiff (ref. 3.5)) that the energy of a linear simple harmonic oscillator can only assume the values

$$E_n = (n + \tfrac{1}{2})\hbar\omega \tag{3.16}$$

where n is a positive integer. The probability that the simple harmonic oscillator has an energy E_n is

$$p(E_n) = \frac{\exp(-E_n/k_B T)}{\sum_{n=0}^{\infty} \exp(-E_n/k_B T)} \tag{3.17}$$

and the mean energy is

$$\bar{E} = \sum_{n=0}^{\infty} E_n p(E_n) \tag{3.18}$$

Making use of results such as

$$\sum_{n=0}^{\infty} e^{-nx} = (1 - e^{-x})^{-1} \quad \text{and} \quad \sum_{n=0}^{\infty} ne^{-nx} = e^x(e^x - 1)^{-2}$$

we arrive at

$$\bar{E} = \hbar\omega \left\{ \frac{1}{2} + \frac{1}{\exp\left(\dfrac{\hbar\omega}{k_B T}\right) - 1} \right\} \tag{3.19}$$

which we can also write as

$$\bar{E} = \hbar\omega(\bar{n} + \tfrac{1}{2}) \tag{3.20}$$

Applying this result to the one-dimensional crystal it is evident that

$$\bar{U}(q) = \hbar\omega(q)(\bar{n}(q) + \tfrac{1}{2}) \tag{3.21}$$

where

$$\bar{n}(q) = \left\{ \exp\left(\frac{\hbar\omega(q)}{k_B T}\right) - 1 \right\}^{-1} \tag{3.22}$$

and

$$\bar{U} = \sum_q \bar{U}(q) \tag{3.23}$$

It is easy to show using the binomial expansion that when $k_B T \gg \hbar\,\omega(q)$ we recover the classical result, Equ. 3.13.

One quantum of lattice vibrational energy $\hbar\omega(q)$ is often referred to as a phonon, but the word really carries deeper implications. Let us anticipate a result of Chapter 4, that corresponding to the waves of different wavenumber q in the one-dimensional crystal there are travelling waves of wavevector **q** and frequency $\omega(\mathbf{q})$ in a three-dimensional crystal. Consider cavity radiation (black body radiation) in an enclosure at temperature T. Only waves which satisfy the boundary conditions are allowed, and each such mode can be regarded as an independent oscillator whose energy is quantized in the same way as we have just been considering. On the other hand the system can also be correctly described in terms of photons which are particles obeying Bose–Einstein statistics and whose number is not conserved. By these two different routes one can arrive at the same expression for the energy of the system and for the distribution of energy over the different wavelengths, the well known Planck distribution. The photon is the particle of the electromagnetic field, the phonon is the analogous particle of the field of the mechanical energy of a crystal. The analogy is not complete however; cavity radiation involves a continuum, the vibrating crystal does not. The frequency of a photon is proportional to the magnitude of its wavevector **k**, and the photon carries momentum $\hbar\mathbf{k}$ which makes itself apparent when the radiation is absorbed or reflected. The phonon corresponding to wavevector **q** does not however carry a momentum $\hbar\mathbf{q}$. This can be appreciated from the fact that **q** can be increased by a vector of the reciprocal lattice without $\omega(\mathbf{q})$ or the atomic displacements being changed. In other words, as we have already seen for the one-dimensional crystal, the dispersion relation for phonons is a periodic function of wavevector, not a linear one as for photons. However, as we shall see later, in the interaction of phonons with one another and with other particles the quantity $\hbar\mathbf{q}$ has properties very reminiscent of momentum; it is often referred to as 'crystal momentum' or 'pseudo-momentum', especially when considering the conservation relations in scattering processes that are discussed in §5.2.

3.3 Longitudinal and transverse modes of vibration

As a first step towards a three-dimensional crystal we can allow the atoms of the one-dimensional crystal to be displaced in any one of three perpendicular directions, u_3 being a longitudinal displacement, u_1 and u_2 transverse displacements. The force constants for the two transverse displacements will be equal, but generally there will be a different force constant for longitudinal displacements. For every wavenumber q, therefore, there will be three independent modes of vibration, two of which have the same frequency. Other conditions such as the periodic relation between frequency and wavenumber, the number of allowed values of q, etc., are obviously not affected. The dispersion curves are shown in Fig. 3.3. We say that the dispersion relation has three branches, two of which in this case are degenerate in frequency.

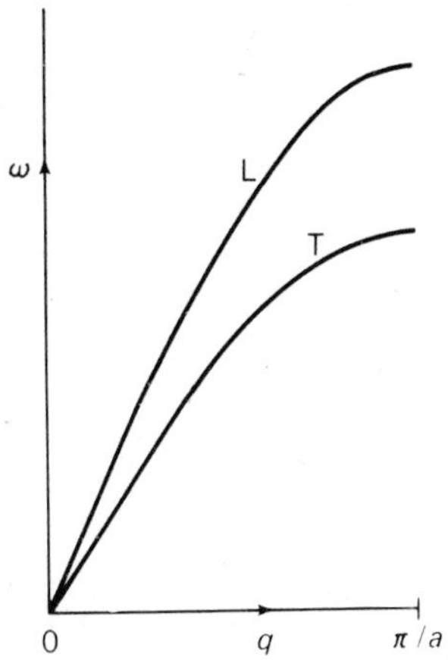

Figure 3.3 The dispersion relation showing transverse and longitudinal branches.

3.4 Long range forces

As an indication of what we can expect when interatomic forces extend further than to nearest neighbours, let us consider again the one-dimensional crystal containing one atom per unit cell, with only longitudinal displacements allowed. Represent the interaction between atoms at r_l and r_{l+p}, which are separated by a distance pa, by a spring of force constant f_p. The equation of motion is then

$$m\frac{\partial^2 u_l}{\partial t^2} = \sum_p f_p\{u_{l+p} - u_l + u_{l-p} - u_l\} \tag{3.24}$$

Substituting the solution given by Equ. 3.2, one finds for the dispersion relation

$$m\omega^2(q) = 2\sum_p f_p(1 - \cos pqa) \tag{3.25}$$

In all other respects the situation remains the same as was found in §3.1. Only the shape of the dispersion curve is altered; the greater the range of the interatomic force the greater the number of terms in the Fourier series (Equ. 3.25) for $\omega^2(q)$; that is, there will be more 'structure' in the dispersion curve. The dispersion curve shown in Fig. 3.4 is appropriate to $f_1 = f, f_2 = \frac{1}{2}f$, $f_3 = f_4 = \text{etc.} = 0$.

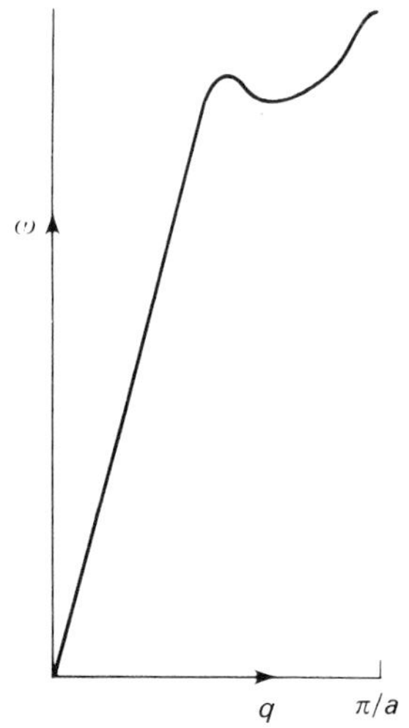

Figure 3.4 The dispersion relation when interatomic forces extend to second nearest neighbours.

3.5 Acoustic and optic modes

Except for crystals of some of the elements, such as copper and sodium, the primitive unit cell contains more than one atom. The trigonal unit cell of NaCl for example contains one Na and one Cl atom. One can understand how the dispersion relation is affected by considering a one-dimensional crystal with two atoms per unit cell. We shall identify the atoms by the letters G and H, since these do not denote particular elements and cannot be confused with any other symbol we have used. The masses are m_G and m_H and their positions in the unit cell r_G and r_H. We could of course choose the origin to make r_G zero if we wished. The force constants on either side of an atom will generally be unequal as shown in Fig. 3.5. Each atom of the crystal is now identified by an integer l

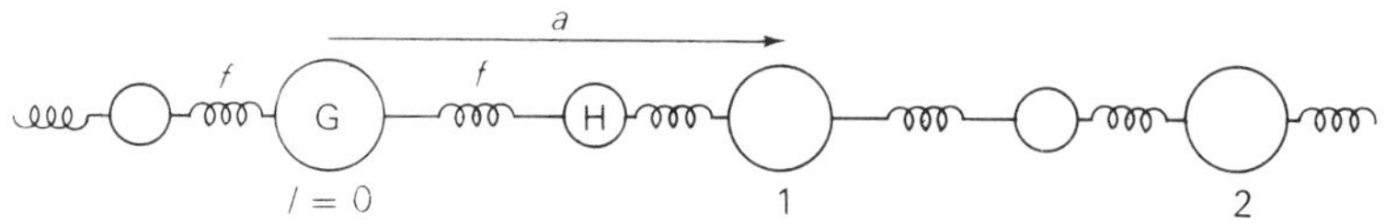

Figure 3.5 A one-dimensional crystal with two atoms per unit cell.

which specifies the unit cell, and a subscript G or H. The position of any atom of type G can be written as

$$r_{lG} = r_l + r_G \tag{3.26}$$

The equations of motion for the atoms in the l'th unit cell, on the simplifying assumption that the interaction is confined to nearest neighbours, are

$$\left.\begin{aligned} m_{\mathrm{G}} \frac{\partial^2 u_{l\mathrm{G}}}{\partial t^2} &= f\{u_{l\mathrm{H}} - u_{l\mathrm{G}}\} + f'\{u_{l-1,H} - u_{l\mathrm{G}}\} \\ m_{\mathrm{H}} \frac{\partial^2 u_{l\mathrm{H}}}{\partial t^2} &= f'\{u_{l+1,\mathrm{G}} - u_{l\mathrm{H}}\} + f\{u_{l\mathrm{G}} - u_{l\mathrm{H}}\} \end{aligned}\right\} \tag{3.27}$$

To solve these equations, or rather to check an assumed solution, we try for the displacement of an atom of type G the equation

$$u_{l\mathrm{G}} = (Nm_{\mathrm{G}})^{-1/2} \sum_q |B(q)|\, |e_{\mathrm{G}}(q)| \cos(qr_{l\mathrm{G}} - \omega(q)t + \alpha(q) + \psi_{\mathrm{G}}(q))$$

We shall find that as before $|B(q)|$ and $\alpha(q)$ remain arbitrary and are determined by the initial conditions but $|e_{\mathrm{G}}(q)|$ and $\psi_{\mathrm{G}}(q)$ are determined by the equations of motion. This is the reason for dividing both amplitude and phase into two parts. It is now more convenient to represent a travelling wave by complex exponentials than by a cosine, since the phase angles $\alpha(q)$ and $\psi_{\mathrm{G}}(q)$ can then be incorporated into complex numbers $B(q)$ and $e_{\mathrm{G}}(q)$ respectively, in the way indicated by Equ. 3.3. We rewrite the last equation as

$$\begin{aligned} u_{l\mathrm{G}} = \tfrac{1}{2}(Nm_{\mathrm{G}})^{-1/2} \sum_q [&B(q)e_{\mathrm{G}}(q) \exp i(qr_{l\mathrm{G}} - \omega(q)t) \\ &+ B^*(q)e^*_{\mathrm{G}}(q) \exp i(-qr_{l\mathrm{G}} + \omega(q)t)] \end{aligned} \tag{3.28}$$

Equ. 3.28 is often written without the factor $\frac{1}{2}$, leaving the second term in the square bracket to be understood. An equation similar to 3.28 is assumed to hold for $u_{l\mathrm{H}}$. Substituting the assumed solutions into the two equations of motion (3.27) and separating out terms which have the same time dependence, one finds after cancelling common factors,

$$\left.\begin{aligned} \omega^2(q)e_{\mathrm{G}}(q) &= D_{\mathrm{GG}}(q)e_{\mathrm{G}}(q) + D_{\mathrm{GH}}(q)e_{\mathrm{H}}(q) \\ \omega^2(q)e_{\mathrm{H}}(q) &= D_{\mathrm{HG}}(q)e_{\mathrm{G}}(q) + D_{\mathrm{HH}}(q)e_{\mathrm{H}}(q) \end{aligned}\right\} \tag{3.29}$$

where

$$\begin{aligned} D_{\mathrm{GG}}(q) &= \frac{f+f'}{m_{\mathrm{G}}}, \quad D_{\mathrm{HH}}(q) = \frac{f+f'}{m_{\mathrm{H}}} \\ -D_{\mathrm{GH}}(q) &= \frac{f}{\sqrt{m_{\mathrm{G}}m_{\mathrm{H}}}} \exp iq(r_{l\mathrm{H}} - r_{l\mathrm{G}}) + \frac{f'}{\sqrt{m_{\mathrm{G}}m_{\mathrm{H}}}} \exp iq(r_{l-1,H} - r_{l\mathrm{G}}) \\ -D_{\mathrm{HG}}(q) &= \frac{f}{\sqrt{m_{\mathrm{G}}m_{\mathrm{H}}}} \exp iq(r_{l\mathrm{G}} - r_{l\mathrm{H}}) + \frac{f'}{\sqrt{m_{\mathrm{G}}m_{\mathrm{H}}}} \exp iq(r_{l+1,\mathrm{G}} - r_{l\mathrm{H}}) \end{aligned} \tag{3.30}$$

$D_{GG}(q)$ and $D_{HH}(q)$ are constants independent of q but are written in this way for the sake of symmetry in Equations 3.29, and also because they are in fact functions of q when the interactions between atoms extend beyond nearest neighbours. We also note that since

$$r_{l-1,H} - r_{lG} = -r_{l+1,G} + r_{lH}$$

it follows that $D_{HG}(q) = D^*_{GH}(q)$. The reader is advised to work through the derivation in this section. It looks more complicated than it is, and is being considered in some detail so that we need not labour through it in three dimensions.

Equations 3.29 can be written in matrix notation as

$$\omega^2(q)\,\mathbf{e}(q) = \mathbf{D}(q)\,\mathbf{e}(q) \tag{3.31}$$

where

$$\mathbf{e}(q) = \begin{pmatrix} e_G(q) \\ e_H(q) \end{pmatrix} \quad \text{is a column matrix}$$

and

$$\mathbf{D}(q) = \begin{pmatrix} D_{GG}(q) & D_{GH}(q) \\ D_{HG}(q) & D_{HH}(q) \end{pmatrix} \quad \text{is a square Hermitian matrix.}$$

The condition for Equations 3.29 to have a non-trivial solution is

$$\begin{vmatrix} D_{GG}(q) - \omega^2(q) & D_{GH}(q) \\ D_{HG}(q) & D_{HH}(q) - \omega^2(q) \end{vmatrix} = 0 \tag{3.32}$$

(See for example, Margenau and Murphy (ref. 3.6).) On multiplying out the determinant we obtain a quadratic equation for $\omega^2(q)$ so that for every q there are now two independent travelling waves of quite different frequencies. We distinguish these by indices $j = 1$ and $j = 2$. For example when $m_G = m_H = m$ the frequencies are given by

$$\left.\begin{aligned} m\omega_1^2(q) &= f + f' - \{(f+f')^2 - 2ff'\sin^2(\tfrac{1}{2}qa)\}^{1/2} \\ &\text{and} \\ m\omega_2^2(q) &= f + f' + \{(f+f')^2 - 2ff'\sin^2(\tfrac{1}{2}qa)\}^{1/2} \end{aligned}\right\} \tag{3.33}$$

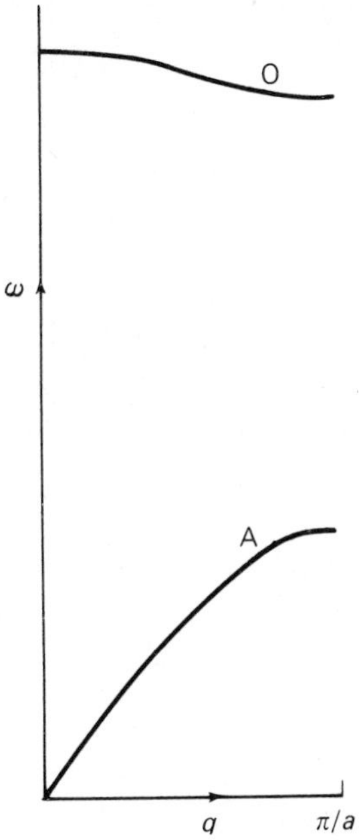

Figure 3.6 The dispersion relation for a diatomic one-dimensional crystal, showing acoustic and optic branches. The ratio of f to f' in Equations 3.33 was taken as 2·00.

This dispersion relation is shown in Fig. 3.6. The branches distinguished by $j = 1$ and $j = 2$ are known as the acoustic and optic branches respectively. For small values of q the frequency of the former is proportional to q while that of the latter is independent of q. The two branches are also distinguished by having different patterns of movement of the atoms. The pattern of movement in an acoustic mode for example is obtained by substituting $\omega_1^2(q)$ in Equations 3.29 and solving for $e_{G1}(q)$ and $e_{H1}(q)$, or rather their ratio since absolute magnitudes are not determined by these equations. It is however convenient to impose the normalizing condition

$$|e_{Gj}(q)|^2 + |e_{Hj}(q)|^2 = 1 \qquad (3.34)$$

In the special case where the masses are equal, one finds that in an acoustic mode ($j = 1$)

$$\frac{e_{H1}(q)}{e_{G1}(q)} = \frac{z(q)}{|z(q)|}$$

while in an optic mode ($j = 2$)

$$\frac{e_{H2}(q)}{e_{G2}(q)} = -\frac{z(q)}{|z(q)|}$$

where

$$z(q) = \exp iq(r_G - r_H)\{f + f' \exp(iqa)\} \tag{3.35}$$

For $q \to 0$, the two types of atom move with one another in an acoustic mode, against one another in an optic mode.† This is illustrated in Fig. 3.7.

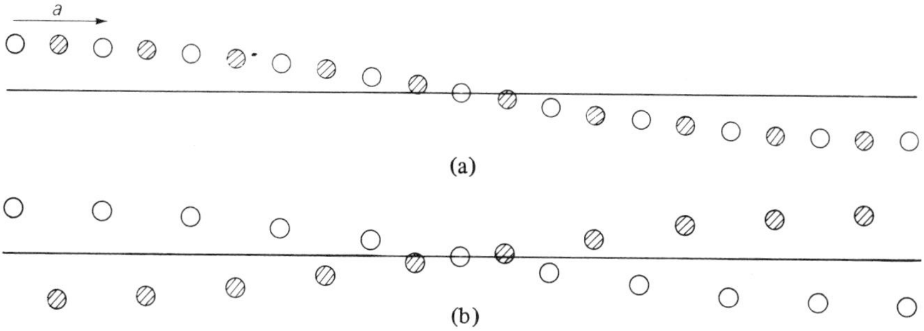

Figure 3.7 The pattern of displacement of the atoms of a diatomic one-dimensional crystal in (a) an acoustic mode and (b) an optic mode, both of the same long wavelength.

In the language of matrix algebra, the $\omega_j^2(q)$ are the eigenvalues of the matrix $\mathbf{D}(q)$ and the $e_j(q)$ are the corresponding eigenvectors. We can see now why the amplitude in Equ. 3.28 was written $B(q)e_G(q)$. The equations of motion do not determine $B_j(q)$ for a particular mode; this quantity is determined by the initial conditions. Whatever the energy and initial phase of the wave, the ratio $e_{Gj}(q)$ to $e_{Hj}(q)$, that is the pattern of displacement of the two types of atom, is always the same for that particular mode of vibration. The general solution for the displacement of an atom is

$$u_{lG} = (Nm_G)^{-1/2} \sum_{qj} |B_j(q)|\, |e_{Gj}(q)| \cos(qr_{lG} - \omega_j(q)t + \alpha_j(q) + \psi_{Gj}(q)) \tag{3.36}$$

Use of periodic boundary conditions shows that the number of independent values of q is the same as was considered in §3.1. There are still N distributed uniformly in the Brillouin zone. The effect of doubling the number of atoms in

† The simplest types of crystal possessing such modes are ionic crystals with the Na^+Cl^- and Cs^+Cl^- structures, and the strong coupling of such modes in those crystals to dipole electromagnetic radiation gives them their name.

the unit cell is not to change the allowed values of q but to increase the number of modes of vibration per value of q to two.

When there are n atoms G,H,I . . . in each unit cell, the derivation is very similar. The equation (3.31) then involves an $n \times n$ matrix $\mathbf{D}(q)$, the elements of which depend on the wavenumber q and on the atomic masses, positions and force constants. The eigenvalues of the matrix (sometimes called the dynamical matrix) are the squared frequencies $\omega_j^2(q)$ with $j = 1, 2, \ldots, n$ and the corresponding eigenvectors determine the pattern of atomic motion in the mode of vibration specified by q and j. Only one of the branches, which we can label $j = 1$, has the characteristic feature of an acoustic mode, namely $\omega_1(q)$ proportional to q when q is small.

References and suggestions for further reading

3.1 Coulson, C. A. 1949, *Waves*, Oliver and Boyd, Edinburgh.

3.2 Cochran, W. and Cowley, R. A. 1967, *Handbook of Physics*, **25/2a**, 59, Springer, Berlin.

3.3 Maradudin, A. A., Montroll, E., and Weiss, G. 1963, *Theory of Lattice Dynamics in the Harmonic Approximation*, Suppl. 3 of *Solid State Physics*, Academic Press, London.

3.4 Jensen, H. H. 1964, in *Phonons and Phonon Interactions*, Ed. T. A. Bak, Benjamin, New York.

3.5 Schiff, L. I. 1968, *Quantum Mechanics*, McGraw-Hill, New York.

3.6 Margenau, H. and Murphy, G. M. 1953, *The Mathematics of Physics and Chemistry*, Van Nostrand, New York.

4

Dynamics of Three-dimensional Crystals

4.1 Force constants

We have not called this chapter 'the dynamics of real crystals' because the theory outlined at this stage is only a first approximation. In particular we continue to use the harmonic approximation, that the forces acting on an atom when atoms are displaced from equilibrium are proportional to the first power of the displacements. This will be correct only when the displacements are very small compared with interatomic distances and we therefore expect the theory to work best at low temperatures. In Chapter 8 we shall see that it usually serves as a first approximation, with anharmonic (i.e., non-linear) effects treated as a perturbation, over the whole temperature range.

The concept of a force constant needs further examination before it can be applied in three dimensions. To simplify matters as far as possible, (and what follows is a simplification of the accepted theory), let us consider a crystal with one atom per unit cell, and let the potential energy of two atoms separated by $\mathbf{r}$ be $\phi(\mathbf{r})$. If these atoms were in isolation, we would expect the potential to depend only on the distance r, and indeed we might expect this still to be true for atoms in a crystal of argon or sodium, but not for atoms in a covalently bonded crystal such as diamond. The potential must be a minimum when summed over all atoms in the equilibrium configuration, and must not change when the crystal is translated or rotated. At this stage we shall use xyz for the Cartesian components of $\mathbf{r}$. Let us fix our attention on the atom in cell l. The force on it in the x direction due to the atom in cell l' is

$$F_{l_x} = \left(\frac{\partial \phi(\mathbf{r})}{\partial x}\right)_{\mathbf{r}=\mathbf{r}_{l'}-\mathbf{r}_l} \tag{4.1}$$

This quantity is not necessarily zero, but since the atom at $\mathbf{r}_l$ is in equilibrium, the resultant force from all other atoms must be zero, so that

$$0 = \sum_{l'} \left(\frac{\partial \phi(\mathbf{r})}{\partial x}\right)_{\mathbf{r}=\mathbf{r}_{l'}-\mathbf{r}_l} \tag{4.2}$$

If now the atom at $\mathbf{r}_{l'}$ is displaced by $u_{l'y}$ in the y direction, the change in the x component of the resultant force on that at $\mathbf{r}_l$ is

$$\Delta F_{lx} = u_{l'y}\left(\frac{\partial^2 \phi(r)}{\partial x\, \partial y}\right)_{r=r_{l'}-r_l} \tag{4.3}$$

The linear relation between force and displacement defines a force constant which we write as $f_{ll',xy}$. The form of Equ. 4.3 shows that there can be six different force constants between any two atoms. This number is often reduced by symmetry. For example consider two nearest neighbour atoms in a

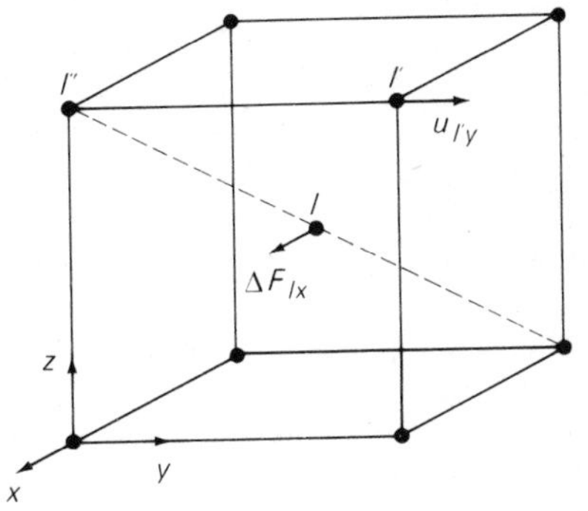

Figure 4.1 Displacement of the atom at site l' produces a force, whose x component is shown, on the atom at site l.

sodium or a potassium crystal (Fig. 4.1). By visualizing the forces produced on atom l by small displacements of atom l' it can be concluded that

$$f_{ll',xx} = f_{ll',yy} = f_{ll',zz} \quad \text{and}$$
$$f_{ll',xy} = f_{ll',xz} = f_{ll',yz} \tag{4.4}$$

so that there are only two independent force constants. Not only that, but the interaction between atom l and all its near neighbours is determined by these two force constants. For example it is readily seen by referring to Fig. 4.1 that $f_{ll',xx} = f_{ll'',xx}$. In more complicated structures the relations between force constants imposed by the space group symmetry can be found using group theory, but we shall not take the subject further.

When the interatomic potential depends only on r there can be only two independent force constants between any pair of atoms. To see this, take the x axis to be the line of centres. By symmetry, only the second derivatives

$$\frac{\partial^2 \phi(\mathbf{r})}{\partial x^2} \quad \text{and} \quad \frac{\partial^2 \phi(\mathbf{r})}{\partial y^2} = \frac{\partial^2 \phi(\mathbf{r})}{\partial z^2} \quad \text{can be non-zero.}$$

These are called the radial and tangential force constants. The components of the force constant tensor in a rotated coordinate system are expressible in terms of these two.

4.2 The normal modes

We can now write down the equations of motion, still for simplicity confining our attention to simple structures with one atom per unit cell. The first is, for atom l,

$$m\frac{\partial^2 u_{l\mathrm{x}}}{\partial t^2} = \sum_{l'} (f_{ll',\mathrm{xx}}\, u_{l'\mathrm{x}} + f_{ll',\mathrm{xy}}\, u_{l'\mathrm{y}} + f_{ll',\mathrm{xz}}\, u_{l'\mathrm{z}}) \tag{4.5}$$

and there are two similar equations for motion in the y and z directions. By analogy with the discussion in §3.5 we anticipate a solution

$$u_{l\mathrm{x}} = (Nm)^{-1/2} \sum_{\mathbf{q}} |B(\mathbf{q})|\, e_{\mathrm{x}}(\mathbf{q}) \cos(\mathbf{q}\,.\,\mathbf{r}_l - \omega(\mathbf{q})t + \alpha(\mathbf{q})) \tag{4.6}$$

The plane wave is now specified by a wavevector $\mathbf{q}$ which determines both the wavelength and the direction of propagation in space. For simple structures the eigenvector $e_{\mathrm{x}}(\mathbf{q})$ turns out to have no imaginary component and we have anticipated this to simplify the notation. The solutions for $u_{l\mathrm{y}}$ and $u_{l\mathrm{z}}$ are similar and involve $e_{\mathrm{y}}(\mathbf{q})$ and $e_{\mathrm{z}}(\mathbf{q})$. On substituting these assumed solutions into the equations of motion we arrive at the result corresponding to Equ. 3.29, which we write as

$$\omega^2(\mathbf{q})e_{\mathrm{x}}(\mathbf{q}) = D_{\mathrm{xx}}(\mathbf{q})e_{\mathrm{x}}(\mathbf{q}) + D_{\mathrm{xy}}(\mathbf{q})e_{\mathrm{y}}(\mathbf{q}) + D_{\mathrm{xz}}(\mathbf{q})e_{\mathrm{z}}(\mathbf{q}) \tag{4.7}$$

with two similar equations.

As before the elements of the 3 x 3 dynamical matrix are determined, for a particular value of $\mathbf{q}$, by the force constants and the geometry of the crystal structure. For example,

$$-mD_{\mathrm{xy}}(\mathbf{q}) = \sum_{l'} f_{ll',\mathrm{xy}} \exp(i\mathbf{q}\,.\,(\mathbf{r}_{l'} - \mathbf{r}_l)) \tag{4.8}$$

The eigenvalues of the matrix are the squared frequencies $\omega_{\mathrm{j}}^2(\mathbf{q})$ for $j = 1,2,3$ and corresponding to each frequency or branch of the spectrum of lattice vibrations there are three eigenvectors $e_{\mathrm{xj}}(\mathbf{q})$, $e_{\mathrm{yj}}(\mathbf{q})$, $e_{\mathrm{zj}}(\mathbf{q})$ which determine the pattern of atomic displacement in this particular mode. Since these satisfy the normalizing condition

$$e_{\mathrm{xj}}^2(\mathbf{q}) + e_{\mathrm{yj}}^2(\mathbf{q}) + e_{\mathrm{zj}}^2(\mathbf{q}) = 1 \tag{4.9}$$

we can introduce a unit vector $\mathbf{e}_j(\mathbf{q})$ for each mode, the polarization vector. It can be shown that $\mathbf{e}_1(\mathbf{q})$, $\mathbf{e}_2(\mathbf{q})$ and $\mathbf{e}_3(\mathbf{q})$ are mutually perpendicular. In general their orientation relative to **q** depends on the force constants but when **q** is in certain symmetry directions such as [100], [110] and [111] in a cubic

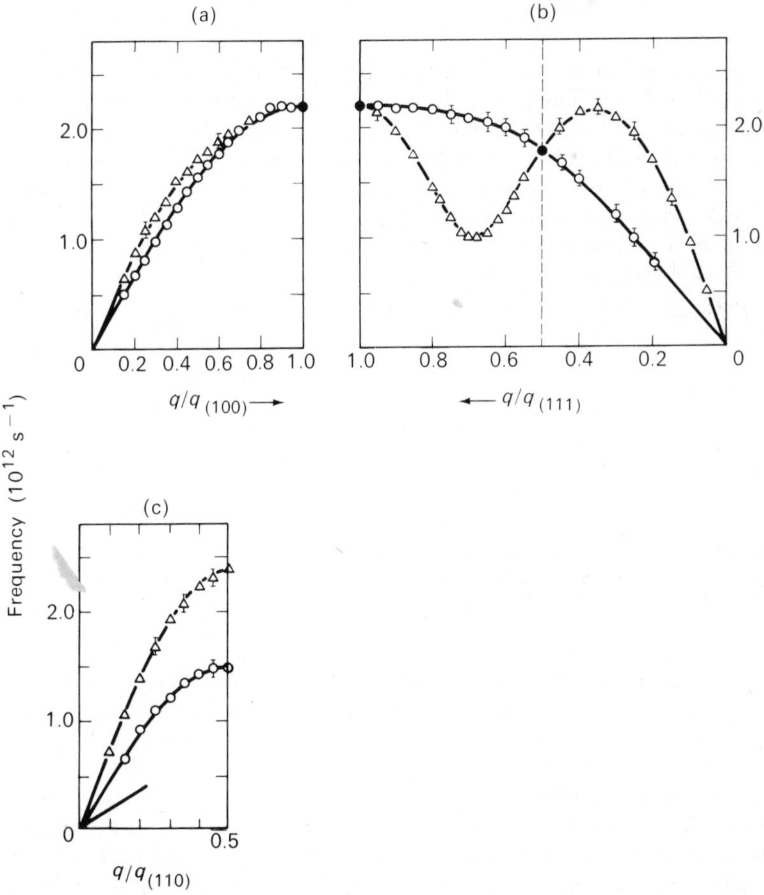

Figure 4.2 The measured frequency as a function of wavevector **q** for normal modes of vibration of potassium at 9 K. (a) **q** along the line joining the origin and the point (100) of the reciprocal lattice, where **q** has the magnitude $2\pi/a$. (b) **q** along the line joining the origin and the point (111) of the reciprocal lattice where **q** has the magnitude $\sqrt{3}\pi/a$. The origin is on the right of the diagram. (c) **q** along the line joining the origin and the point (110) where **q** has the magnitude $\sqrt{2}\pi/a$. Measurements were not made for the lowest transverse branch but the initial slope is shown. (After Cowley, Woods and Dolling, 1966, *Phys. Rev.* **150**, 487.)

crystal of simple structure, the modes are constrained by symmetry to be purely transverse or purely longitudinal. The general displacement of an atom when all modes are simultaneously present can now be written as

$$\mathbf{u}_l = (Nm)^{-1/2} \sum_{\mathbf{q}j} |B_j(\mathbf{q})|\, \mathbf{e}_j(\mathbf{q}) \cos(\mathbf{q}\,.\,\mathbf{r}_l - \omega_j(\mathbf{q})t + \alpha_j(\mathbf{q})) \tag{4.10}$$

By analogy with the discussion in §3.3, we expect all three branches to be acoustic, and this is found to be the case. As before, everything repeats with the periodicity of the reciprocal lattice. This can be seen from the fact that an element such as $D_{xy}(\mathbf{q})$ of the dynamical matrix (see Equ. 4.8) always involves $\mathbf{q}$ as $\exp(i\mathbf{q}\,.\,(\mathbf{r}_{l'} - \mathbf{r}_l))$. When $\mathbf{q}$ is increased by a reciprocal lattice vector $\mathbf{K}_h$ the complex exponential is unchanged, because $\mathbf{K}_h\,.\,(\mathbf{r}_{l'} - \mathbf{r}_l)$ is 2π times an integer (Equ. 2.8), Consequently the dynamical matrix is unchanged.

The phonon dispersion curves for potassium have been determined experimentally for $\mathbf{q}$ in each of three directions, and are shown in Fig. 4.2. The

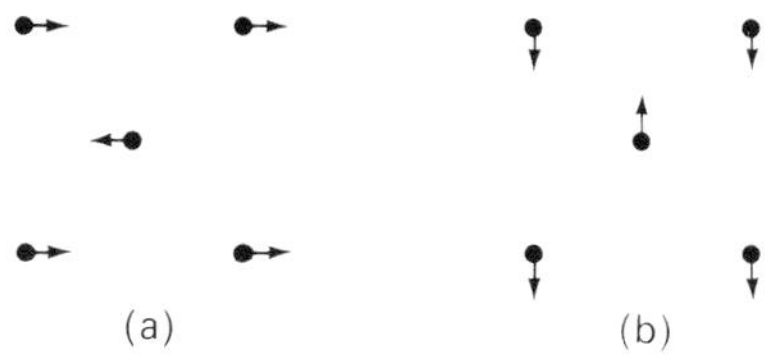

Figure 4.3 Displacements of neighbouring atoms in potassium in (a) longitudinal and (b) transverse modes for which $\mathbf{q}$ is at the point (100), which is the right hand boundary of Fig. 4.2a. The direction of $\mathbf{q}$ is horizontal in the diagram.

experimental method will be described in Chapter 5. We recall from Chapter 2 that (100) and (111) are not genuine points of the reciprocal lattice appropriate to the b.c.c. structure of potassium; (110) is however a reciprocal lattice point and the zone boundary in this direction is met at $(\frac{1}{2}\frac{1}{2}0)$. This explains the range over which the curves are drawn. For $\mathbf{q}$ in the [100] direction, the two transverse branches have the same frequency, lying somewhat below that of the longitudinal branch, except at the zone boundary where they coincide. The explanation of this latter feature can be seen by referring to Fig. 4.3, which illustrates the displacements of atoms for $\mathbf{q}$ having components $2\pi/a$, 0, 0 and polarization (i) longitudinal and (ii) transverse. The two patterns can however be brought into coincidence by a rotation of 90° and cannot therefore correspond to different frequencies of vibration. If we label the branches 1, 2 and 3 in order of increasing frequency with $\mathbf{q}$ in the [110] direction, $\mathbf{e}_1(\mathbf{q})$ is along $[1\bar{1}0]$, $\mathbf{e}_2(\mathbf{q})$ is along [001] and $\mathbf{e}_3(\mathbf{q})$ is along [110]. The two transverse modes have the same frequency when $\mathbf{q}$ is parallel to [111].

4.3 Boundary conditions

In the previous section we were content to draw the $\omega_j(\mathbf{q})$ as continuous curves. As before, the allowed values of $\mathbf{q}$ are determined by the boundary conditions. There is less obvious justification for using periodic boundary conditions in three dimensions than there was in Chapter 3 where we could visualize one end of a chain being joined to the other. It can however be shown, and is indeed intuitively obvious, that the calculation of bulk properties such as the specific heat cannot be influenced by the choice of boundary conditions, as long as the crystal is sufficiently large that atoms at the surface comprise a negligible fraction of the total. We therefore imagine a crystal of indefinite extent to be made up of contiguous identical blocks, each measuring $N_1\mathbf{a}_1, N_2\mathbf{a}_2, N_3\mathbf{a}_3$ as in §2.5. We need not assume, as in the other sections of this chapter, that the crystal is cubic but we shall make the simplifying assumption that the axes are mutually perpendicular. From the condition that $\mathbf{u}_l$ is to be the same for atoms separated by a translation $N_i\mathbf{a}_i$, with $i = 1, 2$ or 3, or a sum of such translations, it can be shown using Equ. 4.6 that allowed values of $\mathbf{q}$ are separated by $2\pi/N_1a_1$, $2\pi/N_2a_2$ and $2\pi/N_3a_3$ in the $\mathbf{b}_1$, $\mathbf{b}_2$ and $\mathbf{b}_3$ directions of the reciprocal lattice respectively. Thus a volume $(2\pi)^3/V$ of reciprocal space surrounds the end of each vector $\mathbf{q}$, where $V = N_1N_2N_3v$ is the volume of the crystal. It follows that there are just $N = N_1N_2N_3$ different values of $\mathbf{q}$ in the Brillouin zone, since the latter occupies a volume $(2\pi)^3/v$ of reciprocal space (see Chapter 2). These are the results which we expect by analogy with the one-dimensional crystal. For N at all large, it is justifiable to regard $\mathbf{q}$ as uniformly and continuously distributed in reciprocal space.

4.4 Multi-atom crystals

When there is more than one atom in the unit cell, the algebra appears even more forbidding. We can however guess the outcome from the consideration of a one-dimensional multi-atom crystal, §3.5. When there are n atoms per primitive unit cell, there are $3n$ different values of j, that is, $3n$ branches of the dispersion relation. Of these, three are acoustic branches and the remaining $3n - 3$ are optic branches. The interatomic force constants and the geometry of the structure determine the elements of the $3n \times 3n$ dynamical matrix. The eigenvalues and eigenvectors of this matrix for a particular value of $\mathbf{q}$ determine the frequency and the pattern of movement of the atoms for each mode of vibration. Fig. 4.4 shows dispersion curves for KBr with $\mathbf{q}$ in symmetry directions. For $\mathbf{q}$ parallel to [100] or [111] the two transverse acoustic modes have the same frequency by symmetry, as have the two transverse optic modes. For $\mathbf{q}$ parallel to [110] the frequencies of the transverse acoustic and transverse optic modes which are polarized parallel to $[1\bar{1}0]$ were not determined, for a reason explained in §5.3. Thus only four, and not six, branches are shown in each section of Fig. 4.4.

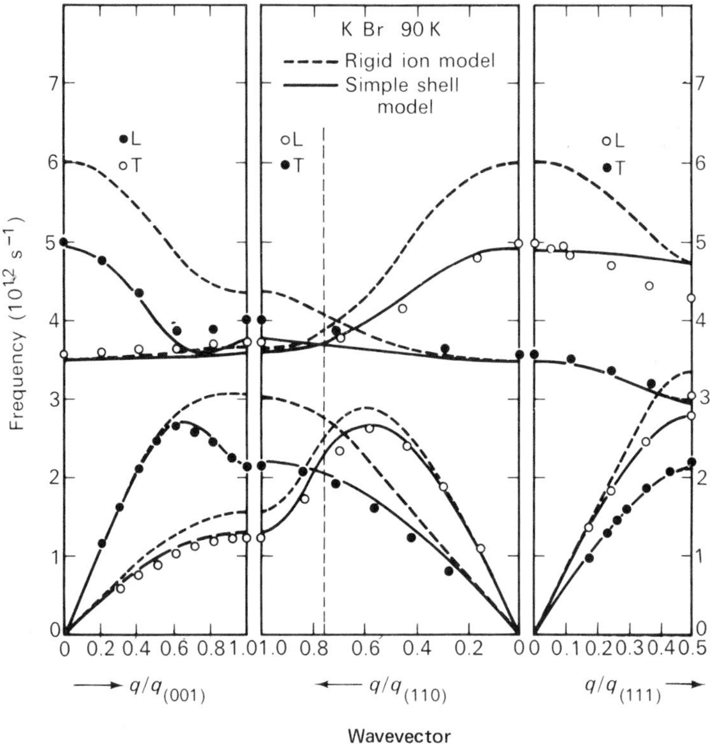

Figure 4.4 The frequency as a function of wavevector **q** for **q** in each of three symmetry directions in a crystal of KBr at 90 K. Measured frequencies are compared with calculated results based on the rigid ion model (dotted lines) and shell model (full lines). (After Woods, Cochran and Brockhouse, 1960, *Phys. Rev.* **119**, 980.)

4.5 Energy and quantization

There is little to add to what we found in Chapter 3. When we make use of the normalizing condition for the eigenvectors (Equ. 4.9; there is a similar result for multi-atom crystals), the energy of a mode is found to be

$$\bar{U}_j(\mathbf{q}) = \tfrac{1}{2}\omega_j^2(\mathbf{q})|B_j(\mathbf{q})|^2 \tag{4.11}$$

We have written earlier of $B_j(\mathbf{q})$ being 'determined by the initial conditions', but this is not very meaningful in relation to an atomic system, and the closest we can come to determining $B_j(\mathbf{q})$ in a crystal in thermal equilibrium is through the equation

$$\tfrac{1}{2}\omega_j^2(\mathbf{q})|B_j(\mathbf{q})|^2 = (\bar{n}_j(\mathbf{q}) + \tfrac{1}{2})\hbar\,\omega_j(\mathbf{q}) \tag{4.12}$$

where $n_j(\mathbf{q})$ is given by Equ. 3.22.

The concept of phonons in a crystal has already been introduced in §3.2.

The topic of this chapter is discussed in much greater detail in the book by Born and Huang, cited at the end of Chapter 1.

5

Determining Phonon Dispersion Curves

5.1 Neutron scattering

Certain bulk properties of a crystal, such as the specific heat, are determined by an average over the whole phonon spectrum. This means that while the specific heat can be predicted when the $\omega_j(\mathbf{q})$ relation is known, the converse does not hold. Generally it is through the interaction of radiation with the crystal that its dynamical properties can be most effectively investigated. Radiation in the optical or infra-red region of the electromagnetic spectrum interacts mainly with optic modes of long wavelength, and we shall consider this topic in Chapter 7. If we wish to obtain detailed information about lattice vibrations for which the value of $\mathbf{q}$ is anywhere in the Brillouin zone we must study their interaction with radiation of wavelength comparable with interatomic distances and having an energy quantum $\hbar\omega$ comparable with phonon energies. X-rays satisfy the first of these conditions, but not the second since $\hbar\omega \sim 10^4$ eV for X-rays while $\hbar\omega_j(\mathbf{q}) \sim 0{\cdot}01$ eV for phonons. However, a beam of neutrons having a velocity $v_n = 2$ km sec^{-1} has a wavevector $\mathbf{k} = m_n\mathbf{v}_n/\hbar$ or wavelength $2\pi/k = 2{\cdot}0$ Å and the energy quantum is $\frac{1}{2}m_n v_n^2 = 0{\cdot}02$ eV. The scattering of a beam of mono-energetic neutrons of about this wavelength provides the most powerful method of studying lattice vibrations and measuring phonon energies. Such experiments are experiments in neutron spectroscopy rather than in neutron diffraction since they involve the measurement of the change in wavelength (or energy, or velocity) of the scattered neutrons, and not merely the intensity of scattering in a particular direction.

There are two mechanisms by which an atom can scatter neutrons in the energy range with which we are concerned. An atom with uncompensated electron spin, or with an electron configuration which otherwise gives it a magnetic moment, scatters by interaction with the magnetic moment of the neutron. While this provides an important method of investigating magnetism on an atomic scale, we shall not be concerned with it here. The nucleus of an atom scatters neutrons much as if it were a hard sphere of radius 10^{-12} cm in order of magnitude, and it is this nuclear scattering with which we shall be concerned. We saw in Chapter 2 that the scattering of X-rays by a single atom, fixed in position, involved the atomic scattering factor $f(K)$. The corresponding quantity for neutron scattering is the scattering length b, which is simpler than $f(K)$ in that it is independent of neutron wavelength or the angle of scattering. Thus the amplitude of the neutron beam scattered by a crystal which has one atom per unit cell can be written

$$A_c(\mathbf{K}) = b \sum_l \exp(i\mathbf{K}\,.\,\mathbf{r}_l) \tag{5.1}$$

(compare Equ. 2.19), where $\mathbf{K} = \mathbf{k}_1 - \mathbf{k}_0$ is the difference between the wave-vectors of the incident and scattered neutron beams. We assume that all the atoms are fixed in position, so that the scattering is elastic, with $k_1 = k_0$. As before, Equ. 5.1 reduces to

$$A_c(\mathbf{K}) = \frac{(2\pi)^3}{v} b \sum_h \delta(\mathbf{K} - \mathbf{K}_h) \tag{5.2}$$

so that the scattered amplitude is zero unless $\mathbf{K}$ coincides with a vector $\mathbf{K}_h$ of the reciprocal lattice. When there are several atoms per unit cell, b is replaced in Equ. 5.2 by the structure factor

$$F(\mathbf{K}_h) = \sum_{\kappa=1}^{n} b_\kappa \exp(i\mathbf{K}_h\,.\,\mathbf{r}_\kappa) \tag{5.3}$$

where b_κ is the scattering length of an atom of type κ. Experimental measurements of the Bragg intensities give the values of $|F(\mathbf{K}_h)|^2$, from which the crystal structure can be determined. This technique is referred to as neutron diffraction, to distinguish it from neutron spectroscopy, and is discussed in detail in ref. 2.4.

In the above brief introduction we have of course swept some difficulties under the carpet, and for a fuller discussion the reader is referred to ref. 2.4 and to 'Neutron Diffraction' by G. Bacon (ref. 5.1). In particular we have considered only coherent scattering. Different isotopes of the same element may not have the same scattering length. For example, for Fe^{54}, $b = 0{\cdot}42 \times 10^{-12}$ cm and for Fe^{56}, $b = 1{\cdot}01 \times 10^{-12}$ cm. Thus in considering scattering by a crystal of iron, b cannot be factored out of the sum in Equ. 5.1 and instead we must write

$$A_c(\mathbf{K}) = \sum_l b_l \exp(i\mathbf{K}\,.\,\mathbf{r}_l) = \bar{b} \sum_l \exp(i\mathbf{K}\,.\,\mathbf{r}_l) + \sum_l (b_l - \bar{b}) \exp(i\mathbf{K}\,.\,\mathbf{r}_l) \tag{5.4}$$

where $\bar{b}$ is the average scattering length. The first term involving $\bar{b}$, obviously 'peaks' at points of the reciprocal lattice and determines the Bragg intensity. The mean value of the second term is zero, but its mean square value is $N(\overline{b^2} - \bar{b}^2)$. This corresponds to a uniform intensity of scattering in all directions, proportional to the number of atoms, and is referred to as scattering by 'isotope incoherence'. An additional source of incoherent scattering arises from the fact that a nucleus of non-zero spin has two different scattering lengths, b_+ and b_-, depending on the relative orientations of nuclear and neutron spin. These may be quite different, for example for hydrogen $b_+ = 1{\cdot}04 \times 10^{-12}$ cm and $b_- = -4{\cdot}7 \times 10^{-12}$ cm. There is therefore always a relatively high background of 'spin incoherent scattering' from hydrogenous materials, so much so that they are unsuitable for the type of experiment which we discuss in a later section. In what follows it is the coherent scattering length $\bar{b}_\kappa$ which is involved for each element; however we write it simply as b_κ and ignore incoherent scattering.

5.2 Theory of neutron spectroscopy

Displacement of the nuclei from their equilibrium positions also results in scattering in directions which do not satisfy the condition $\mathbf{K} = \mathbf{K}_\text{h}$. Let us first consider the effect of static displacements,

$$\mathbf{u}_l = (Nm)^{-1/2}|B_\text{j}(\mathbf{q})|\, \mathbf{e}_\text{j}(\mathbf{q}) \cos(\mathbf{q}\,.\,\mathbf{r}_l + \alpha_\text{j}(\mathbf{q})) \tag{5.5}$$

corresponding to a single wave of wavevector $\mathbf{q}$ and polarization vector $\mathbf{e}_\text{j}(\mathbf{q})$. The amplitude of the scattered beam is now

$$\begin{aligned} A_\text{c}(\mathbf{K}) &= b \sum_l \exp(i\mathbf{K}\,.\,(\mathbf{r}_l + \mathbf{u}_l)) \\ &\triangleq b \sum_l \exp(i\mathbf{K}\,.\,\mathbf{r}_l) + ib\mathbf{K}\,.\,\sum_l \mathbf{u}_l \exp(i\mathbf{K}\,.\,\mathbf{r}_l) \end{aligned} \tag{5.6}$$

The first term evidently corresponds to Bragg scattering. The second may be written

$$A_\text{c}'(\mathbf{K}) = \tfrac{1}{2}(Nm)^{-1/2}|B_\text{j}(\mathbf{q})|(\mathbf{K}\,.\,\mathbf{e}_\text{j}(\mathbf{q}))\, b\,[\sum_l \exp i(\mathbf{K}+\mathbf{q})\,.\,\mathbf{r}_l + \sum_l \exp i(\mathbf{K}-\mathbf{q})\,.\,\mathbf{r}_l] \tag{5.7}$$

on making use of Equ. 5.5. The two sums in square brackets are the same as were met before in §2.5. They peak at $\mathbf{K} + \mathbf{q} = \mathbf{K}_\text{h}$ and $\mathbf{K} - \mathbf{q} = \mathbf{K}_\text{h}$ respectively. Thus scattered intensity appears in additional directions defined by $\mathbf{K} = \mathbf{K}_\text{k} \pm \mathbf{q}$. The effect is precisely analogous to the faint additional spectra (ghosts) which are produced by a periodic variation in the spacing of the rulings of a diffraction grating (see Fig. 5.1).

Now suppose that waves corresponding to all possible wavevectors and states of polarization j are simultaneously present, the amplitude $|B_\text{j}(\mathbf{q})|$ varying only slowly with $\mathbf{q}$ for a particular value of j, but the phase angle $\alpha_\text{j}(\mathbf{q})$ being quite random. The effect can be considered in two stages. First of all the large

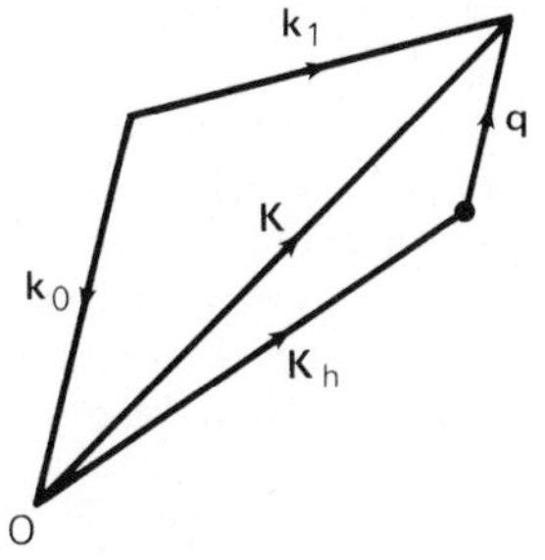

Figure 5.1 The relation between the scattering vector **K**, vector to a reciprocal lattice point $\mathbf{K}_h$, and the phonon wavevector **q**.

number of small independent displacements gives each nucleus an isotropic Gaussian probability distribution about its mean position $\mathbf{r}_l$. This 'spreading out' of the nucleus† has the same effect on scattering as the extended nature of the electron cloud has on X-ray scattering, interference between the radiation scattered from different parts of the distribution or cloud causing the effective scattering length to decrease with increasing K. The effect was first considered for X-ray scattering by Debye and later by Waller who showed that in the present context b is replaced by $b \exp(-w)$ where

$$w = \tfrac{1}{2}K^2 \overline{u^2} \tag{5.8}$$

and $\overline{u^2}$ is the mean square displacement in a particular direction. The second effect of having waves of all possible wavevectors present is of course to give scattering in all directions. On carrying through the detailed calculation, which consists in evaluating $|A_c'(K)|^2$ from Equ. 5.7 and summing the contributions from adjacent modes, it is found that the intensity is determined by

$$S_1(\mathbf{K}) = \left(\frac{N}{2m}\right) b^2 \exp(-2w) \sum_j (\mathbf{K} \cdot \mathbf{e}_j(\mathbf{q}))^2 \, |B_j(\mathbf{q})|^2 \tag{5.9}$$

where the sum is over the three branches of the spectrum. If we now assume that although the displacements are static, the amplitudes are those appropriate to thermal excitation, we can use Equ. 4.12 to obtain

$$S_1(\mathbf{K}) = Nb^2 \exp(-2w) \sum_j \frac{(\mathbf{K} \cdot \mathbf{e}_j(\mathbf{q}))^2 (\bar{n}_j(\mathbf{q}) + \frac{1}{2})\hbar\omega_j(\mathbf{q})}{m\omega_j^2(\mathbf{q})} \tag{5.10}$$

† The range in space of this distribution, unlike the effective extent of the nucleus itself, is not so very much less than the internuclear distance and the neutron wavelength.

As before **K** and **q** are related by the condition that their sum is $\mathbf{K}_h$, a vector of the reciprocal lattice.

The result given by Equ. 5.10 is in fact correct for X-ray scattering, when b is replaced by $f(K)$. The change in frequency of the X-ray photons scattered by atoms in motion is relatively so small that no appreciable error is made by taking the displacements to be static. A neutron beam however is appreciably shifted in frequency (energy) in scattering by the moving atoms. For convenience we introduce a frequency ω such that $\hbar\omega$ is the energy change of scattered neutrons,

$$\hbar\omega = \tfrac{1}{2}m_n(v_{0n}^2 - v_{1n}^2) = \frac{\hbar^2}{2m_n}(k_0^2 - k_1^2) \tag{5.11}$$

The condition of conservation of energy in the scattering process is

$$\hbar\omega = \frac{\hbar^2}{2m_n}(k_0^2 - k_1^2) = \pm\, h\omega_j(\mathbf{q}) \tag{5.12}$$

In other words the crystal exchanges one quantum of lattice vibrational energy, one phonon, with the neutron beam. To take account of this we have to define a scattering cross section such that $S_1(\mathbf{K}\omega)$ determines the probability that neutrons will be scattered into a small solid angle about the direction $\mathbf{k}_1$ with energy change between $\hbar\omega$ and $\hbar(\omega + d\omega)$. The final result is

$$S_1(\mathbf{K}\omega) = Nb^2 \exp(-2w)\sum_j \frac{(\mathbf{K}\,.\,\mathbf{e}_j(\mathbf{q}))^2\hbar\omega_j(\mathbf{q})}{m\omega_j^2(\mathbf{q})}$$
$$[(\bar{n}_j(\mathbf{q}) + 1)\pi\delta(\omega - \omega_j(\mathbf{q})) + \bar{n}_j(\mathbf{q})\pi\delta(\omega + \omega_j(\mathbf{q}))] \tag{5.13}$$

There is thus always a greater probability that neutrons will be scattered with energy loss ($\omega > 0$) than with energy gain ($\omega < 0$), and at low temperatures where $\bar{n}_j(\mathbf{q})$ tends to zero (Equ. 3.22) only scattering with neutron energy loss is possible. On integrating Equ. 5.13 over frequency, Equ. 5.10 is obtained:

$$S_1(\mathbf{K}) = \int S_1(\mathbf{K}\omega)d\omega/2\pi \tag{5.14}$$

In this notation incidentally the intensity of Bragg scattering is determined by

$$S_0(\mathbf{K}\omega) = Nb^2 \exp(-2w)\frac{(2\pi)^3}{v}\sum_h \delta(\mathbf{K} - \mathbf{K}_h)2\pi\delta(\omega) \tag{5.15}$$

which expresses the condition on **K** which we had before, plus the fact that the scattering is elastic.

To summarize the two most important results, one-phonon scattering changes the wavevector of the neutrons from $\mathbf{k}_0$ to $\mathbf{k}_1$ such that

$$\left.\begin{aligned} \mathbf{K} &\equiv \mathbf{k}_1 - \mathbf{k}_0 = \mathbf{K}_h \pm \mathbf{q} \quad \text{and} \\ \hbar\omega &\equiv \frac{\hbar^2}{2m_n}(k_0^2 - k_1^2) = \pm\,\hbar\omega_j(q) \end{aligned}\right\} \tag{5.16}$$

The second equation expresses conservation of energy. The first might be thought to express conservation of momentum, but while $\hbar\mathbf{K}$ is certainly the change of neutron momentum and $\hbar(\mathbf{q} + \mathbf{K}_h)$ that of the crystal, it is not correct to make a further division and identify $\hbar\mathbf{q}$ as literally the phonon momentum. The presence of the factor $\mathbf{K}\,.\,\mathbf{e}_j(\mathbf{q})$ in the formula for the intensity shows that it is the motion of the atoms in the direction of $\mathbf{K}$ that is effective in changing the neutron momentum. As we remarked in §3.2, $\hbar\mathbf{q}$ is sometimes called the pseudo-momentum of the phonon.

5.3 Experimental neutron spectroscopy

In principle, by determining the change in momentum and energy of scattered neutrons, and applying Equations 5.16, $\omega_j(\mathbf{q})$ can be determined. A number of different experimental techniques have been developed, but we shall describe only one of them. In Fig. 5.2, a beam of neutrons which have come into thermal equilibrium with the material of the 'moderator' of a reactor, and thus have energies in the desired range, falls on a single crystal referred to as the monochromator. This crystal, often of germanium, lead or graphite, Bragg reflects a narrow range of wavelengths thus producing a beam having a definite wavevector $\mathbf{k}_0$ and energy $\hbar^2 k_0^2/2m_n$ which passes through a collimating system and falls on the specimen under investigation, which is in a known orientation. The beam scattered in a particular direction is defined by a collimating system which thus defines the direction of $\mathbf{k}_1$. The magnitude of k_1 and the energy of the scattered neutrons are determined by finding the angle at which they are Bragg reflected from a set of planes of a third single crystal, the analyser. This apparatus is referred to as a triple-axis spectrometer.

By varying $\mathbf{k}_0$ and/or $\mathbf{k}_1$ so that $\mathbf{K}$ and therefore $\mathbf{q}$ remains constant but $\hbar\omega = \hbar^2/2m_n\,(k_0^2 - k_1^2)$ runs through a range of values, the frequency (energy) spectrum of neutrons which have been scattered by phonons of a known wavevector can be investigated. According to Equ. 5.13 this spectrum consists of three infinitely sharp peaks on either side of $\omega = 0$ corresponding to scattering by three modes of different polarization, with neutron energy loss or energy gain. In practice of course the peaks are broadened by instrumental factors, and also for a more fundamental reason which we consider in Chapter 8, but the maxima occur at $\pm\omega_j(\mathbf{q})$. It is sometimes possible to utilize the factor $\mathbf{K}\,.\,\mathbf{e}_j(\mathbf{q})$ which occurs in Equ. 5.13 to eliminate scattering by all except one mode of

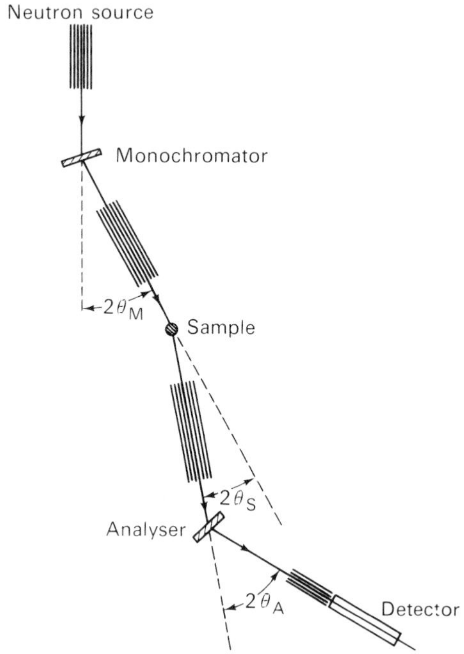

Figure 5.2 Schematic representation of a triple-axis neutron spectrometer. (After Cooper and Nathans, 1967, *Acta Cryst.* **23**, 357.)

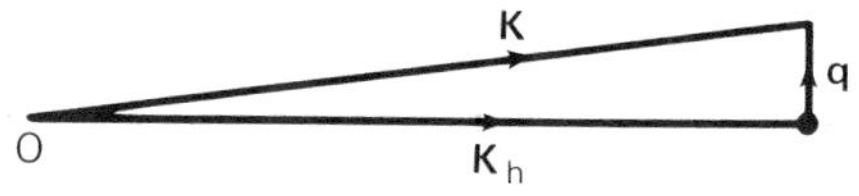

Figure 5.3 Illustrates a situation favourable to scattering by a transverse mode of wavevector **q**. The reciprocal lattice point at $\mathbf{K}_h$ is (100).

known polarization. For example Fig. 5.3 illustrates a situation in which the wavevector **q** is drawn from a point of the reciprocal lattice in the [010] direction so as to satisfy the condition $\mathbf{K} - \mathbf{q} = \mathbf{K}_h$. A wave propagating in this direction in a cubic crystal has purely transverse or longitudinal polarization, and the factor $\mathbf{K} \cdot \mathbf{e}_j(\mathbf{q})$ ensures that only the mode with $\mathbf{e}_j(\mathbf{q})$ parallel to [100] can scatter with appreciable intensity. Thus modes of different polarization can be distinguished. A typical neutron spectrum is shown in Fig. 5.4.

The formulae quoted in §5.2 apply for a crystal with only one atom per unit cell. Similar results apply in the general case, and in particular the sum over j in a modified expression for $S_1(\mathbf{K}\omega)$ is then over all $3n$ branches of the phonon spectrum. Typical results for $\omega_j(\mathbf{q})$ are shown as Fig. 4.2 for K, a monatomic crystal, and as Fig. 4.4 for KBr, a diatomic crystal. It should be noted that for **q** parallel to [110] the transverse branches with polarization in the $[1\bar{1}0]$ direction were not determined since the crystals were oriented with this direction perpendicular to the plane of the scattering vector **K**.

Measurements are usually made with the wavevector **q** in symmetry directions but can in principle be made for any value of **q**.

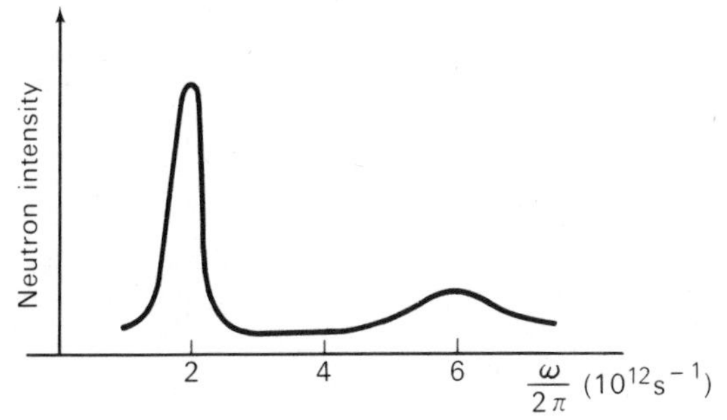

Figure 5.4 A neutron spectrum in which the two peaks correspond to scattering by an acoustic and an optic mode respectively.

5.4 Interpretation of phonon dispersion curves

The curves determined by experiment are mainly of interest for the possibility they provide of testing various models of interatomic forces.† As examples we shall outline the interpretations that have been given for the simplest metallic, ionic and covalent crystals.

It was found that the measured frequencies for potassium could be reproduced within experimental error by assuming radial and tangential force constants (§4.1) extending from an atom to its nearest neighbours out to the fifth nearest. The ten force constants involved were determined by the method of least squares and are shown in Table 5.1. The curves calculated using these parameters are the full lines in Fig. 4.2. These force constants are not very meaningful in themselves, however we saw in §4.1 that the radial force constant is equal to the second derivative of the interatomic potential, $d^2\phi(r)/dr^2$, and it can be shown that the tangential force constant is equal to $(1/r)(d\phi(r)/dr)$. Although knowledge of the first and second derivatives for five values of r tells us comparatively little about $\phi(r)$, it proves possible to account for the force constants in terms of Coulomb interaction between the ions in potassium and

† They also provide the description of the dynamics of the real crystal which should be used in, for example, discussions of electron-phonon interactions in specific materials.

an indirect action between them through the conduction electrons, which has the effect of screening the Coulomb interaction beyond about fifth neighbours.

Table 5.1 Radial and tangential force constants between an atom and its five nearest neighbours in potassium. The units are dyne cm^{-1} (10^5 dyne = 1 newton).

Neighbour	Radial	Tangential
1	2576	−109
2	432	29
3	−95	12
4	3	−4
5	15	2

The most obvious interaction in an alkali halide is again Coulomb interaction between the ions. The cohesive energy is very adequately accounted for in terms of the predominantly attractive electrostatic potential, and a short range repulsive potential between near neighbour ions. The potential between any ion pair is thus assumed to be

$$\phi(r) = A\,\exp(-r/\rho) \pm \frac{e^2}{4\pi\epsilon_0 r} \tag{5.17}$$

where e is the electronic charge. The electrostatic potential energy of the whole crystal is $-N\alpha_M e^2/4\pi\epsilon_0 r_0$ where r_0 is the equilibrium distance between nearest neighbour ions and α_M, the Madelung constant, is a number depending only on the geometry of the crystal structure, (see for example Kittel (ref. 5.2)). From Equ. 5.17 and the discussion of force constants in §4.1 it is clear that the contribution to the radial and tangential force constants between any ion pair, which originates in the Coulomb potential, is completely determined. In calculating the phonon dispersion curves for KBr it was assumed that the repulsive potential extended only to nearest neighbours. Thus its contribution to the force constants can involve only two parameters whether it is specified by the Born-Mayer exponential term in Equ. 5.17 or in some other way. It turns out that this contribution to the tangential force constant between nearest neighbours can be related to the Madelung constant, using the condition that the crystal is in equilibrium under the two opposing interactions. In calculating the phonon dispersion curves for KBr the remaining parameter was chosen to give the observed value for the initial gradient of the longitudinal acoustic mode for which **q** is in the [100] direction. The long range of the Coulomb potential requires the calculations to be made in a way that is more sophisticated mathematically than our discussion involving force constants would suggest, but it is not necessary to go into detail on this point. The calculated curves for KBr are the dotted lines in Fig. 4.4. Considering that only one adjustable parameter is involved, the agreement with the experimental points obtained by neutron spectroscopy is not bad and supports the assumption that KBr is an ionic crystal. There are however obvious discrepancies, particularly for the longitudinal optic modes.

In Chapter 7 we discuss the connection between the dielectric properties of an ionic crystal and its lattice dynamics. The 'rigid ion' model accounts for the dielectric constant by the displacement of positive and negative ions in opposite directions in an electric field. At high frequencies, in the optical region of the electromagnetic spectrum, there is still a contribution to the dielectric constant of an actual crystal although the ions are too massive to respond to the field. This contribution originates in the distortion of the electron distribution; the ions are not rigid but are polarizable, and the outer electrons, particularly of the negative ion in an alkali halide, are the most polarizable. The rigid ion model does not allow for this since the ions are treated as point charges. In the course of lattice vibrations the electric field set up by the displacements of the ions is modified by their electronic polarizability which in turn modifies the force on them and affects the phonon frequencies. A 'shell model' has been used successfully to allow for this. Each ion is taken to be composed of a rigid or non-polarizable core and a charged shell which is bound to the core by an isotropic force constant and can therefore be displaced relative to the core. In the simplest form of the model the short range potential acts exclusively between the shells of nearest neighbour ions. The full lines in Fig. 4.4 are the dispersion curves for KBr calculated in this way, and the agreement with the measurements is much improved.

Germanium was one of the first crystals to be investigated by neutron spectroscopy. It has the diamond structure (see §2.3), with each atom covalently bonded to four nearest neighbours. The expectation that force constants would be appreciable only between nearest neighbour and perhaps next nearest neighbour atoms was not fulfilled; it was found that fifteen independent force constants extending to fifth nearest neighbours were required to give a good fit to the phonon frequencies measured by neutron spectroscopy. Germanium has a relatively high dielectric constant arising entirely from electronic polarizability. If the electron distribution is changed in the course of the lattice vibrations so that the atoms acquire dipole moments, Coulomb interaction becomes almost as important as in an ionic crystal. The situation can be described in terms of a shell model for which the charges on the core and shell are equal and opposite. Although the model has less intuitive appeal when used to represent covalently bonded atoms than it has when used to represent ions with a closed shell electron configuration, it was found in practice to give quite good agreement with the measurements on germanium when force constants between cores and shells extended only to nearest neighbours. Fig. 5.5 shows results for silicon, which has the same crystal structure as germanium. The full lines are calculated for a shell model; to fit the measurements within experimental error force constants extending to second neighbours were necessary.

The theoretical situation is less satisfactory for covalent than for ionic or metallic crystals in that the force constants have to be treated as parameters,

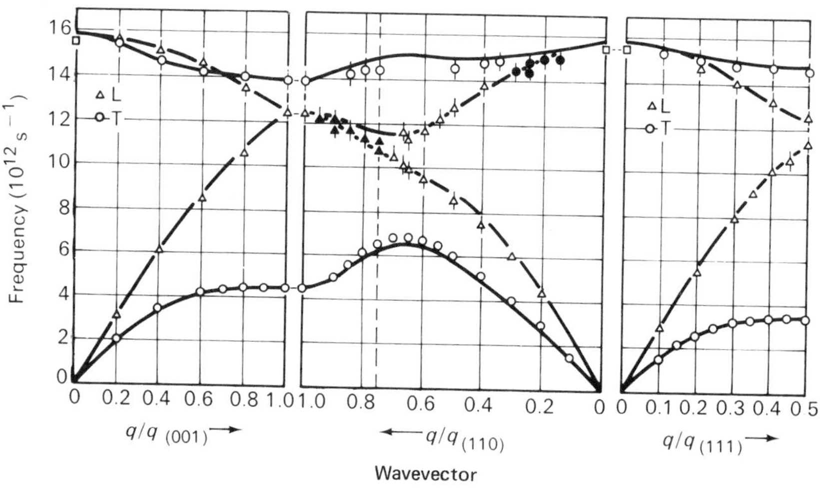

Figure 5.5 Measured frequencies in silicon for the wavevector **q** in each of three symmetry directions, as in Fig. 4.2. The lines give the results of calculations based on a shell model. (After Dolling, article in *Inelastic Scattering of Neutrons in Solids and Liquids*, International Atomic Energy Agency, Vienna, 1963.)

chosen to fit the experimental measurements. As yet little progress has been made in predicting them, or the cohesive energy, from anything approaching first principles.

5.5 Elastic constants

Acoustic modes of long wavelength propagate as in a continuous medium and their frequencies and polarization properties are related to the elastic constants of the crystal. These are defined in the following way. We consider three mutually perpendicular unit vectors $\mathbf{j}_1, \mathbf{j}_2, \mathbf{j}_3$ in the unstrained material. When it is subject to a uniform deformation, these vectors change in length and are no longer mutually perpendicular (Fig. 5.6). For a small deformation, the relation between the new axes and the original can be written

$$\left.\begin{aligned} \mathbf{j}_1' &= (1+\epsilon_{11})\mathbf{j}_1 + \epsilon_{12}\mathbf{j}_2 + \epsilon_{13}\mathbf{j}_3 \\ \mathbf{j}_2' &= \epsilon_{21}\mathbf{j}_1 + (1+\epsilon_{22})\mathbf{j}_2 + \epsilon_{23}\mathbf{j}_3 \\ \mathbf{j}_3' &= \epsilon_{31}\mathbf{j}_1 + \epsilon_{32}\mathbf{j}_2 + (1+\epsilon_{33})\mathbf{j}_3 \end{aligned}\right\} \quad (5.18)$$

It appears that nine quantities are necessary to define the strain at the point in the medium which we are considering. However, Equ 5.18 does not exclude the possibility that the new axes are related to the original system by a rotation, involving no deformation of the medium at all. We therefore define the strains

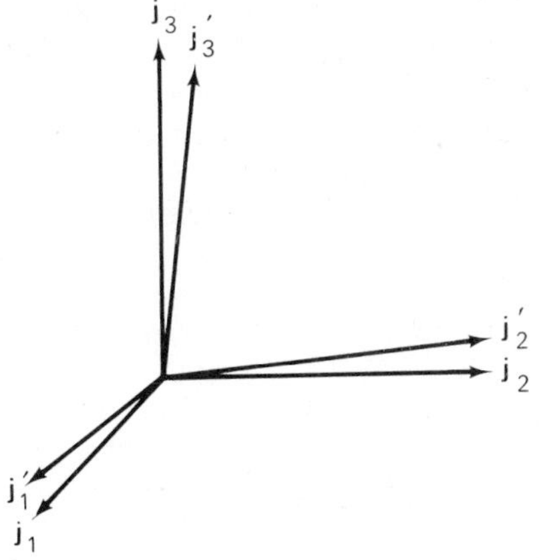

Figure 5.6 Vectors in unstrained and in elastically strained material.

in terms of the length of j_1' etc., and the angles between $\mathbf{j}_1'$ and $\mathbf{j}_2'$, etc. Neglecting quantities of order ϵ_{11}^2, etc. we have from Equ. 5.18 that

$$\left.\begin{aligned} j_1' &= (\mathbf{j}_1' \cdot \mathbf{j}_1')^{1/2} = 1 + \epsilon_{11} \\ \cos\alpha_1 &= \mathbf{j}_2' \cdot \mathbf{j}_3' = \epsilon_{32} + \epsilon_{23} \end{aligned}\right\} \quad (5.19)$$

together with two other similar equations of each kind. For convenience we replace $(\epsilon_{32} + \epsilon_{23})$ by γ_{23}, etc. It is clear that the six quantities ϵ_{11}, ϵ_{22}, ϵ_{33}, γ_{23}, γ_{31}, γ_{12} define the deformation. They are called the strain components.

Now referring to Fig. 5.7, the stress on the cube of material defined by $\mathbf{j}_1, \mathbf{j}_2, \mathbf{j}_3$ can be denoted τ_{11}, τ_{12} etc. The notation is such that the first subscript denotes direction of a force and the second identifies the cube face on which it acts. Thus for example τ_{23} is the force per unit area acting in the

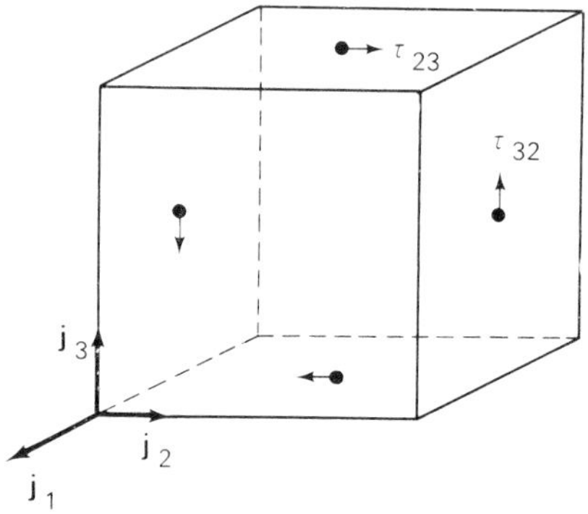

Figure 5.7 Illustrates some of the shear stresses on a cube of material.

direction $\mathbf{j}_2$ across the face perpendicular to $\mathbf{j}_3$ (see Fig. 5.7). Clearly for the unit cube to be in translational and rotational equilibrium there can be only six independent components of stress, and in particular $\tau_{23} = \tau_{32}$ etc. The most general linear relation between components of stress and components of strain can be written

$$\tau_{11} = c_{11\,11}\epsilon_{11} + c_{11\,22}\epsilon_{22} + c_{11\,33}\epsilon_{33} + c_{11\,23}\gamma_{23} + c_{11\,31}\gamma_{31} + c_{11\,12}\gamma_{12} \quad (5.20)$$

plus five similar equations. Some economy can be effected by replacing 11 by 1, 22 by 2, 33 by 3, 23 by 4, 31 by 5 and 12 by 6 so that the representative Equ. 5.20 becomes

$$\tau_{11} = c_{11}\epsilon_{11} + c_{12}\epsilon_{22} + c_{13}\epsilon_{33} + c_{14}\gamma_{23} + \epsilon_{15}\gamma_{31} + c_{16}\gamma_{12} \quad (5.21)$$

The quantities c_{11} etc. are the elastic constants. Since there are a further five equations similar to (5.21) there are apparently 36 independent elastic constants. However it can be shown that, for example, $c_{15} = c_{51}$, so that the maximum number of independent elastic constants is 21. This number is usually much reduced by considerations of symmetry. For a cubic crystal for example the matrix of elastic constant reduces to

$$\begin{matrix} c_{11} & c_{12} & c_{12} & 0 & 0 & 0 \\ c_{12} & c_{11} & c_{12} & 0 & 0 & 0 \\ c_{12} & c_{12} & c_{11} & 0 & 0 & 0 \\ 0 & 0 & 0 & c_{44} & 0 & 0 \\ 0 & 0 & 0 & 0 & c_{44} & 0 \\ 0 & 0 & 0 & 0 & 0 & c_{44} \end{matrix} \quad (5.22)$$

and there are only three independent constants.

Consider a cubic crystal subjected to a uniform (outward) pressure Δp. We then have $\tau_{11} = \tau_{22} = \tau_{33} = \Delta p$, the components of shear stress being zero. Also $\epsilon_{11} = \epsilon_{22} = \epsilon_{33} = \Delta L/L = \frac{1}{3}\Delta V/V$ (where L is the length of a side) and the components of shear strain are zero. Equ. 5.21 together with Equ. 5.22 therefore gives

$$\Delta p = \frac{1}{3}\frac{\Delta V}{V}(c_{11} + 2c_{12}) \tag{5.23}$$

Defining the compressibility, or the reciprocal of the bulk modulus, by

$$\kappa = \frac{1}{V}\frac{\Delta V}{\Delta p} \tag{5.24}$$

we see that $\kappa = 3/(c_{11} + 2c_{12})$ (5.25)

Notice that as we have defined them the elastic constants are measured in Nm^{-2}, and the compressibility in $\mathrm{m}^2\mathrm{N}^{-1}$, where N denotes newtons.

In order to connect elastic constants with the properties of acoustic modes, consider for example a longitudinal wave propagating along a cube axis, taken to be the x or $\mathbf{j}_1$ direction in a cubic crystal. Let a point in the crystal originally at x be displaced to $x + u$, and a point originally at $x + dx$ to $x + dx + u + (\partial u/\partial x)dx$. There is only one strain component, $\epsilon_{11}(x) = \partial u/\partial x$, which we find by considering the change in length of the element dx. The strain is of course a function of x. Now consider a slab of thickness Δx having faces of unit area perpendicular to the x direction (Fig. 5.8). The net force accelerating the slab in the positive x direction is

$$c_{11}\epsilon_{11}(x + \Delta x) - c_{11}\epsilon_{11}(x) = c_{11}\Delta x\frac{\partial^2 u}{\partial x^2} \tag{5.26}$$

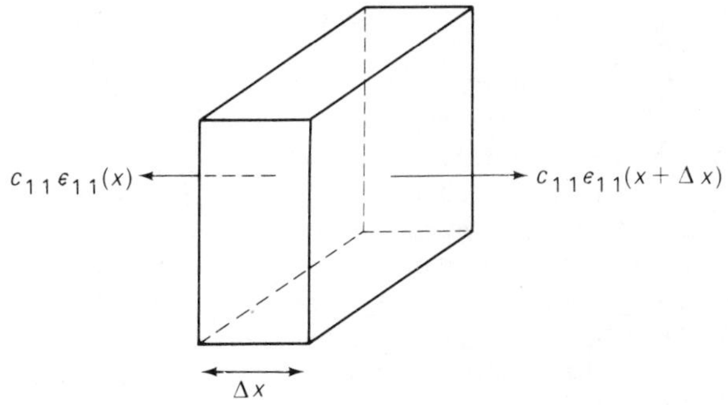

Figure 5.8 Stresses on opposite faces of a slab of thickness Δx.

so that the equation of motion is

$$D\Delta x \frac{\partial^2 u}{\partial t^2} = c_{11}\, \Delta x \frac{\partial^2 u}{\partial x^2} \tag{5.27}$$

where D is the density.

The solution of this equation represents a longitudinal wave travelling with velocity $(c_{11}/D)^{1/2}$. Denoting the longitudinal acoustic branch by the subscript $j = 3$ we thus have, since the velocity is also the ratio of frequency to wave-number, that

$$\omega_3(\mathbf{q}) = q\left(\frac{c_{11}}{D}\right)^{1/2} \tag{5.28}$$

provided that $\mathbf{q}$ is parallel to a cube axis. Similarly we find that for transverse acoustic modes of long wavelength propagating in this direction

$$\omega_1(\mathbf{q}) = \omega_2(\mathbf{q}) = q\left(\frac{c_{44}}{D}\right)^{1/2} \tag{5.29}$$

The velocity in any direction involves a linear combination of the three elastic constants. It is only for waves travelling in the symmetry directions {100}, {110} or {111} however, that the polarization is constrained by symmetry to be purely transverse or purely longitudinal.

The most accurate method of measuring the elastic constants of a crystal consists in 'injecting' into a block of crystal from a quartz transducer, a pulse of ultrasonic waves, of frequency typically 10^7 s^{-1}. The time taken for the pulse to travel across the block and to be reflected from a plane face back to its starting point can be measured electronically. If the waves are longitudinally polarized, and the direction is [100] in a cubic crystal, c_{11} can be immediately deduced. The other elastic constants can be deduced from the velocities of longitudinal and of transverse waves travelling in other directions. Values of the elastic constants of representative cubic crystals, taken from a review article by Huntingdon (ref. 5.3), are given in Table 5.2. The initial slopes of dispersion curves determined by neutron spectroscopy should agree with the values established by the elastic constants, at least in the harmonic approximation, and this is found to be the case in practice, within 2 or 3%. Differences of this order of magnitude are to be expected at ordinary temperatures because of the somewhat different thermal conditions for ultrasonic waves and the waves of generally shorter wavelength accessible to investigation by neutron spectroscopy.

Table 5.2 Elastic constants of some representative cubic crystals. Units are 10^{10} N m^{-2}.

Material	c_{11}	c_{44}	c_{12}
Aluminium	10·82	2·85	6·13
Diamond	10·76	5·76	1·25
Potassium	0·46	0·26	0·37
NaCl	4·87	1·26	1·24

The initial slopes of the acoustic branches, like everything else of the $\omega_j(\mathbf{q})$ relation, are determined by the interatomic forces. They are particularly sensitive to forces of long range. It was shown many years ago by Cauchy that if the forces between atoms are central forces, that is if the interatomic potential $\phi(\mathbf{r})$ depends only on r, and the crystal is in equilibrium under the action of these forces, and furthermore every atom is situated at a centre of symmetry, then there will be relations between the elastic constants additional to those given by crystal symmetry. For a cubic crystal the Cauchy relation is $c_{12} = c_{44}$. The experimental values for most alkali halides satisfy this condition moderately well. When the electronic polarizability of the ions is allowed for by using the simpler versions of the shell model, the calculated elastic constants are the same as for the rigid ion model, and therefore still satisfy the Cauchy relation. However, this is not the case when the theory is based on more elaborate versions of the shell model. In an alkali metal such as sodium, the screened Coulomb interaction between the ions is a central force, but the equilibrium volume of the crystal depends also on the fact that the kinetic energy of the conduction electrons is a function of volume. Thus the electron gas contributes to the compressibility, i.e. to $(c_{11} + 2c_{12})$, but not to c_{44} or to $\frac{1}{2}(c_{11} - c_{12})$ which are the elastic constants involved in shearing deformations. Consequently even the simplest theories which take these points into consideration do not predict $c_{12} = c_{44}$, nor is this relation satisfied by the experimental values. The elastic constants of covalent crystals such as diamond and germanium cannot be expected to satisfy the Cauchy relation for a variety of reasons; to begin with the atoms are not at centres of symmetry.

Because of anharmonic effects, which we discuss in Chapter 9, elastic constants are not independent of temperature, an increase of 10% or more in going from ordinary temperature to 0 K being fairly typical.

References and suggestions for further reading

5.1 Bacon, G. 1962, *Neutron Diffraction*, Clarendon Press. Oxford.

5.2 Kittel, C. 1971, *Introduction to Solid State Physics*, John Wiley and Sons, New York and London.

5.3 Huntingdon, H. B. 1958, *Solid State Physics*, 7, 213, Academic Press, New York.

Several of the topics of this chapter are considered in more detail in *Thermal Neutron Scattering* edited by P. Egelstaff, 1965, Academic Press, New York.

6

Thermal Properties of Crystals

6.1 The specific heat

The theory of the specific heat of a crystal has already been discussed qualitatively in Chapter 1. We recall that in the harmonic approximation the energy of the crystal is the sum of the energies of the independent normal modes. We saw in Chapter 3 that the average energy of an oscillator of frequency ω at temperature T is

$$\bar{E} = \hbar\omega\left(\frac{1}{2} + \frac{1}{\exp(\hbar\omega/k_B T) - 1}\right) \tag{6.1}$$

In Chapter 4 we decided that we knew completely the dynamics of a crystal with n atoms per unit cell when we knew the frequency $\omega_j(\mathbf{q})$ and the pattern of displacement of the atoms (the eigenvectors) for each mode, that is for every value of $\mathbf{q}$ in the first Brillouin zone and for each branch of the dispersion curves, $j = 1, 2, \ldots, 3n$. The energy of the crystal is therefore

$$\begin{aligned} \bar{U} &= \sum_{\mathbf{q}j} \hbar\omega_j(\mathbf{q})\left(\frac{1}{2} + \frac{1}{\exp(\hbar\omega_j(\mathbf{q})/k_B T) - 1}\right) \\ &= \sum_{\mathbf{q}j} \hbar\omega_j(\mathbf{q})[\tfrac{1}{2} + \bar{n}_j(\mathbf{q})] \end{aligned} \tag{6.2}$$

It is interesting to note that the energy depends on the temperature and on the frequencies of the modes only; it is independent of the polarization properties or of the pattern of displacement of the atoms in each mode. A longitudinal acoustic mode and a transverse optic mode for example, have the same energy if their frequencies happen to be the same. We may also note that at absolute zero the crystal still has 'zero-point energy',

$$\bar{U}_0 = \tfrac{1}{2}\hbar\sum_{\mathbf{q}j} \omega_j(\mathbf{q}) \tag{6.3}$$

When the number of unit cells N in the crystal is large, as it must be for us to neglect the effect of the surface and treat the specific heat as a bulk property, the modes are practically continuously distributed over a frequency range from zero to some maximum value ω_m, and we can replace the sum in Equ. 6.2 by an integral. We define a frequency distribution $G(\omega)$ such that $G(\omega)d\omega$ is the number of modes having frequencies in the range ω to $\omega + d\omega$. Equ 6.2 then becomes

$$\bar{U} = \int_0^{\omega_m} \hbar\omega\left(\frac{1}{2} + \frac{1}{\exp(\hbar\omega/k_BT) - 1}\right) G(\omega)d\omega \tag{6.4}$$

Since the total number of normal modes in a crystal of N unit cells with n atoms per unit cell is $3Nn$, we also have that

$$\int_0^{\omega_m} G(\omega)d\omega = 3Nn \tag{6.5}$$

The specific heat is now obtained by differentiating Equ. 6.4 with respect to temperature:

$$C_v = \frac{d\bar{U}}{dT} = k_B \int_0^{\omega_m} \left(\frac{\hbar\omega}{k_BT}\right)^2 \frac{\exp(\hbar\omega/k_BT)G(\omega)d\omega}{[\exp(\hbar\omega/k_BT) - 1]^2} \tag{6.6}$$

The integral can be evaluated numerically when $G(\omega)$ is known, thus the specific heat can be evaluated at any temperature provided the frequency distribution is known.

As an example we consider the frequency distribution for the one-dimensional crystal with forces between nearest neighbour atoms, which we considered in Chapter 3. Equ. 3.4 gives

$$\omega(q) = \omega_m \sin \tfrac{1}{2}qa \tag{6.7}$$

We saw in §3.1 that successive values of q are separated by an interval $2\pi/Na$ (Equ. 3.6). Hence introducing a distribution in wavenumber we have in an obvious notation that

$$G(q) = \frac{Na}{2\pi} \tag{6.8}$$

But

$$2G(q)dq = G(\omega)d\omega \tag{6.9}$$

the factor 2 coming from the fact that we are limiting q to positive values.

Using

$$\frac{dq}{d\omega} = \frac{2}{a(\omega_m^2 - \omega^2)^{1/2}} \tag{6.10}$$

from Equ. 6.7, the frequency distribution is

$$G(\omega) = \frac{2N}{\pi(\omega_m^2 - \omega^2)^{1/2}} \tag{6.11}$$

This distribution is shown in Fig. 6.1, and the specific heat which it gives when substituted in Equ. 6.6 is shown in Fig. 6.2. As would be anticipated from the discussion given in Chapter 1, C_v is zero when $T = 0$ and tends to the value Nk_B for $k_BT \gg \hbar\omega_m$.

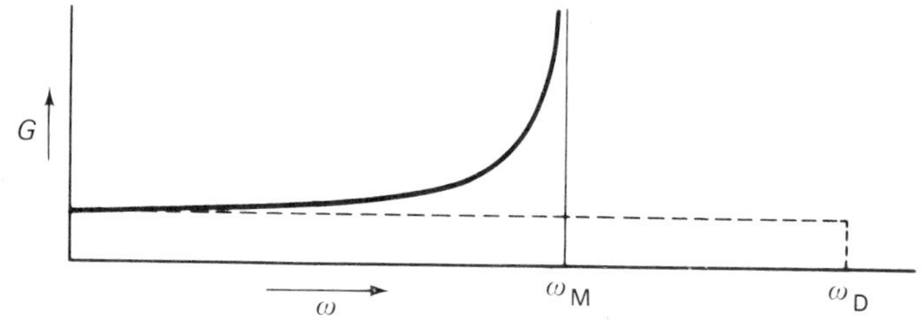

Figure 6.1 The frequency distribution of the normal modes of vibration of a one-dimensional crystal with forces between nearest neighbours (full line). The distribution given by the Debye approximation is also shown (dotted line).

A simpler approximate frequency distribution is obtained by ignoring the dispersion in the $\omega(q)$ relation, replacing the sine curve of Equ. 6.7 by a straight line,

$$\omega(q) = \tfrac{1}{2}\omega_m qa \tag{6.12}$$

as shown in Fig. 6.3. The limiting frequency at the zone boundary is then

$$\omega_D = \omega_m(\pi/2) \tag{6.13}$$

and the corresponding frequency distribution is readily shown to be

$$G_D(\omega) = \frac{2N}{\pi\omega_m} = \frac{N}{\omega_D} \quad \text{for} \quad \omega < \omega_D \tag{6.14}$$

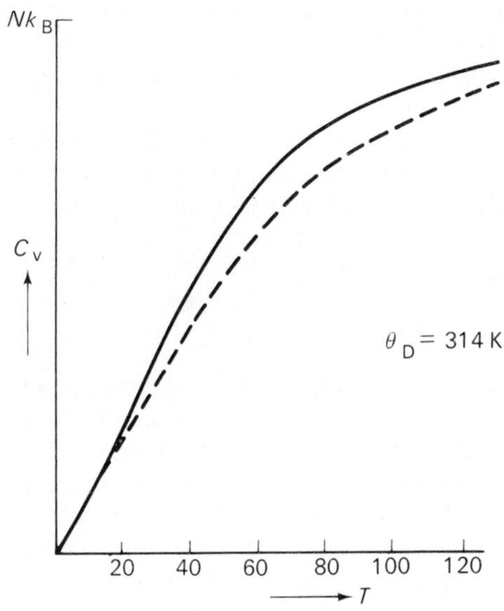

Figure 6.2 The specific heat of a one-dimensional crystal for which the frequency distribution is given in Fig. 6.1. The dotted line gives the result of the Debye approximation.

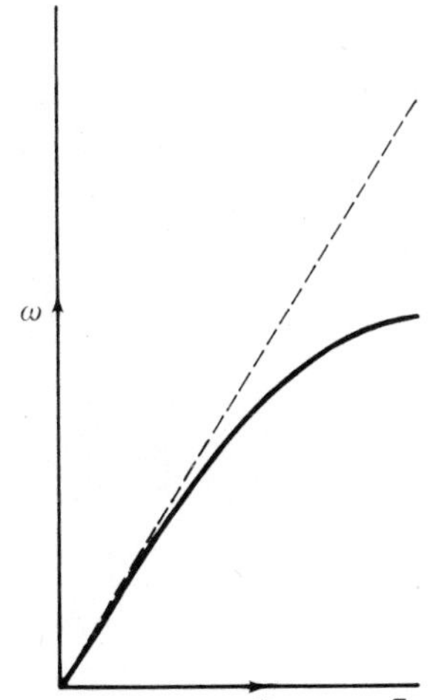

Figure 6.3 The dispersion relation for a one-dimensional crystal (full line) and the corresponding Debye approximation (dotted line).

This approximation to the frequency distribution is the Debye approximation for the one-dimensional crystal. It is shown by the dotted line in Fig. 6.1 and the corresponding specific heat is shown by the dotted line in Fig. 6.2. The approximation is evidently a good one for C_V, in particular it gives *correct* values of C_V at low temperatures. The reason is that at these temperatures, quantization ensures that only modes of low frequency are thermally excited and contribute to C_V, and $G_D(\omega)$ coincides with $G(\omega)$ at the low frequency end of the distribution (see Fig. 6.1).

6.2 The Debye approximation

The frequency distribution $G(\omega)$ for an actual crystal is by no means a simple function of ω, as is apparent from the example shown in Fig. 6.6. The approximation introduced by Debye for $G(\omega)$ however gives values for C_V in good agreement with observation, especially at low temperatures. The approximation consists in ignoring dispersion, just as in the one-dimensional example considered above. Furthermore the various branches of the $\omega_j(\mathbf{q})$ relation are replaced by a single acoustic branch, so that

$$\omega_j(\mathbf{q}) = uq \tag{6.15}$$

where u is a constant, and the Brillouin zone is replaced by a sphere. Since values of $\mathbf{q}$ are uniformly distributed throughout the Brillouin zone (§4.3) it follows that

$$G(q)dq \propto 4\pi q^2 dq \tag{6.16}$$

and therefore from Equ. 6.15,

$$G_D(\omega) = B\omega^2 \tag{6.17}$$

where the subscript D again indicates 'Debye approximation', and B is a constant. Equ. 6.17 will apply up to a maximum frequency corresponding to q being at the surface of the Brillouin zone. This is the Debye frequency ω_D. Equ. 6.17 can be expressed in terms of ω_D using

$$\int_0^{\omega_D} G_D(\omega)d\omega = 3N \tag{6.18}$$

assuming for simplicity that we are dealing with a monatomic crystal, i.e. $n = 1$.

$$\therefore G_D(\omega) = 9N(\omega^2/\omega_D^3) \quad \text{for} \quad \omega < \omega_D \tag{6.19}$$

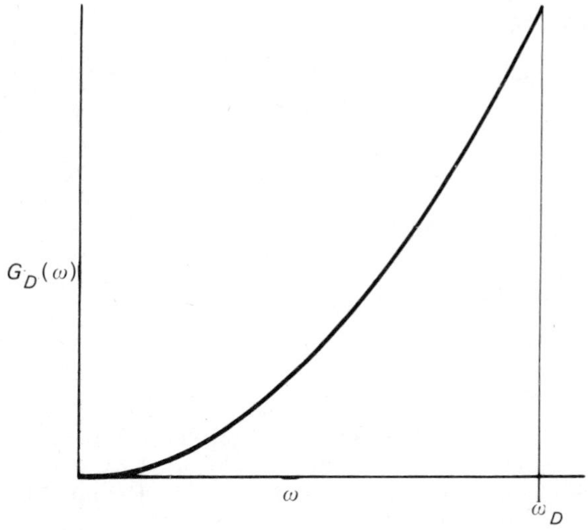

Figure 6.4 The Debye approximation for the frequency distribution of the normal modes of vibration in a three-dimensional crystal.

This is Debye's approximation for the frequency distribution, and is shown in Fig. 6.4. It expresses the dynamical properties of a crystal in terms of a single parameter ω_D, a somewhat drastic approximation. Substituting from Equ. 6.19 in Equ. 6.4, but omitting the zero-point energy since it does not contribute to the specific heat, the thermal energy is given as

$$\bar{U}' = \frac{9N\hbar}{\omega_D^3} \int_0^{\omega_D} \frac{\omega^3 d\omega}{[\exp(\hbar\omega/k_B T) - 1]} \tag{6.20}$$

It is convenient to define a Debye temperature θ_D such that

$$\hbar\omega_D = k_B \theta_D \tag{6.21}$$

and to write

$$x = \frac{\hbar\omega}{k_B T} \quad x_D = \frac{\hbar\omega_D}{k_B T} = \frac{\theta_D}{T} \tag{6.22}$$

Equ. 6.20 then becomes

$$\bar{U}' = 9Nk_BT\left(\frac{T}{\theta_D}\right)^3 \int_0^{x_D} \frac{x^3\,dx}{\exp x - 1} \tag{6.23}$$

and after some algebra the expression for the specific heat is found to be

$$C_v = \frac{d\bar{U}'}{dT} = 9Nk_B\left(\frac{T}{\theta_D}\right)^3 \int_0^{x_D} \frac{x^4 \exp x\,dx}{[\exp x - 1]^2} \tag{6.24}$$

The value of the integral obviously depends only on $x_D = \theta_D/T$, from which it follows that C_v should be a function of T/θ_D, i.e. when expressed as a function of reduced temperature, the specific heat of all materials is predicted to be the same.

This prediction was borne out to a remarkable extent using the experimental measurements available when Debye's theory was first put forward, and for some years thereafter. Fig. 6.5. (reproduced from Becker (ref. 6.1)) shows C_v as given by Equ. 6.24, with experimental points for Pb, Ag and Fe. The Debye temperatures for these materials were taken to be 88, 215 and 453 K respectively. The agreement is good, certainly much better than is given by Einstein's theory of the specific heat which was discussed in Chapter 1.

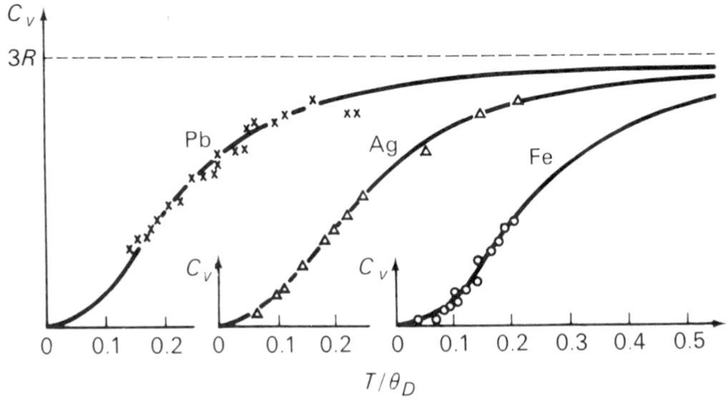

Figure 6.5 The specific heats of Pb, Ag and Fe. Experimental values are compared with those given by Debye's theory (full line). (From Becker, 1967, *Theory of Heat*, 2nd edition, Springer Verlag, Berlin (Fig. 83, p. 249).)

From the definition of x_D it is apparent that for $T \to 0$, $x_D \to \infty$. Thus at low temperatures the limits of the integral in Equ. 6.23 can be taken to be 0 and ∞, and using the result

$$\int_0^\infty \frac{x^3\,dx}{\exp x - 1} = \frac{\pi^4}{15} \tag{6.25}$$

we obtain

$$\bar{U}' = \frac{3Nk_B\pi^4 T^4}{5\theta_D^3} \tag{6.26}$$

and

$$C_v = \frac{12}{5}\pi^4 Nk_B\left(\frac{T}{\theta_D}\right)^3 \tag{6.27}$$

This is the Debye T^3 law for the specific heat of a solid at low temperatures.

The result that the energy is proportional to T^4 is of course also Stefan's law for cavity (black body) radiation. This is not a chance coincidence; the approximation on which Equ. 6.26 is based is that frequency is proportional to wavenumber and this is an exact result for electromagnetic radiation. In other words the Debye frequency distribution is correct for black body radiation, with ω_D infinite.

6.3 'Exact' theory of the specific heat

As more accurate experimental measurements became available, the deficiencies of Debye's theory become apparent, although by this time most physicists had forgotten that Debye's theory involved a drastic approximation, and it was Blackman who pointed out (ref. 6.2) that the correct expression for C_v is Equ. 6.6, with $G(\omega)$ appropriate to the material concerned.

When accurate comparisons are being made, the contribution of the conduction electrons to the specific heat must be taken into account, unless of course the material is an insulator. A very general result for their contribution (see, for example, Mott and Jones (ref. 6.3)) is

$$C_e = \gamma T = \tfrac{2}{3}\pi^2 k_B^2 n(E)_F T \tag{6.28}$$

where $n(E)_F$ is the density of electron states at the Fermi energy. Since the lattice contribution varies approximately as T^3, the electronic contribution becomes the larger at low temperatures. For example, for silver at 1 K they are respectively $0{\cdot}8\,.\,10^{-4}$ and $1{\cdot}6\,.\,10^{-4}$ cal . deg^{-1} mole^{-1}. In the discussion of results for potassium which follows, C_e has been subtracted from the measured specific heat.

We take potassium as an example because the frequency distribution has been determined reliably. As described briefly in §5.4, frequencies $\omega_j(\mathbf{q})$ were measured by neutron spectroscopy for a range of values of **q**, as shown in Fig. 4.2 (Cowley *et al.* (ref. 6.4)). The temperature of the specimen was 9 K. Force constants were found which accounted well for the measured frequencies (Table 5.1), using the theory of lattice dynamics outlined in §4.2. The theory

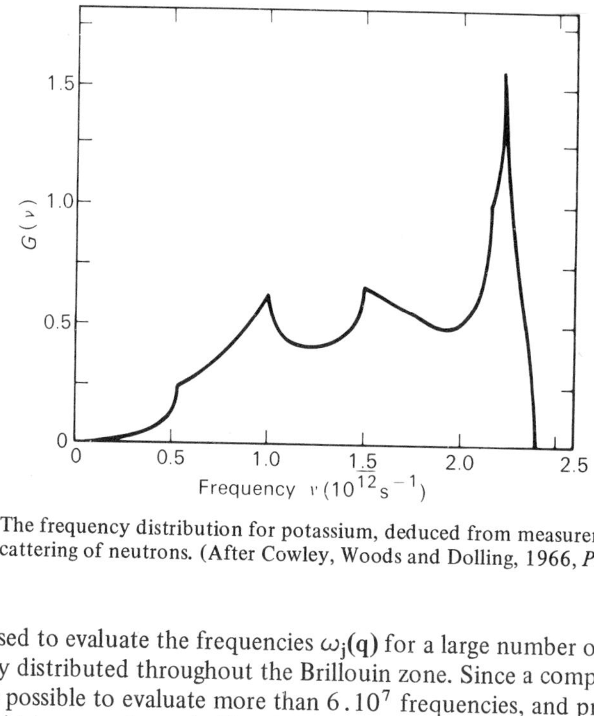

Figure 6.6 The frequency distribution for potassium, deduced from measurements made by inelastic scattering of neutrons. (After Cowley, Woods and Dolling, 1966, *Phys. Rev.* **150**, 487.)

was then used to evaluate the frequencies $\omega_j(\mathbf{q})$ for a large number of values of **q** uniformly distributed throughout the Brillouin zone. Since a computer was used it was possible to evaluate more than $6\,.\,10^7$ frequencies, and present the results as a histogram, shown in Fig. 6.6. Since so many values were available, the histogram gives $G(\omega)$ as a smooth curve, not much resembling a Debye distribution, although it begins by being proportional to ω^2, and has a fairly precipitous cut-off. The Debye theory is so firmly established that in this investigation, as has become customary, the results were presented in terms of a Debye temperature. That is, C_v was evaluated from Equ. 6.6 and Equ. 6.24 was then solved for θ_D at each temperature. θ_D derived from the neutron measurements in this way is of course a function of temperature, whereas in Debye's theory it is a constant. Experimental measurements of C_V were presented in the same way, and the results are compared in Fig. 6.7. The agreement is satisfactory. Most materials yield a curve for $\theta_D(T)$ of the same general form. Evidently the T^3 law, which depends on the constancy of θ_D, can be accurate over a rather small range of temperature, up to no more than about $(1/50)\theta_D$; note that θ_D is still a useful parameter even if it has lost its precision.

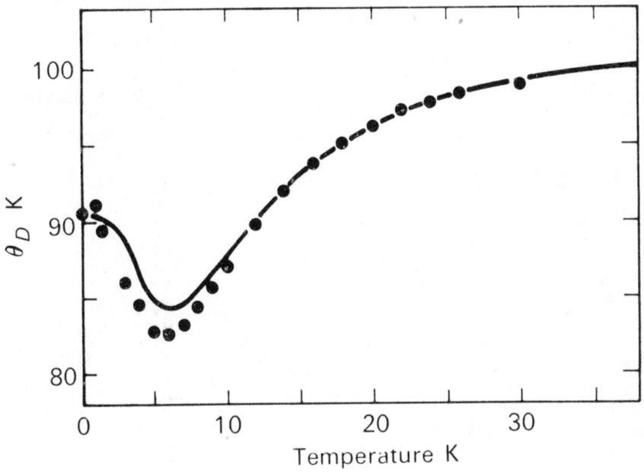

Fig. 6.7 The Debye temperature $\theta_D(T)$ for potassium. Points give the results deduced from measured values of the specific heat; the full line is given by the 'exact' theory, with $G(\omega)$ as given in Fig. 6.6. (After Cowley, Woods and Dolling, 1966, *Phys. Rev.* **150**, 487.)

It should not be forgotten that what we have called the 'exact' theory involves the harmonic approximation. Although we have written the specific heat derived from theory as C_V, we might equally well have written it C_p, because a crystal in which the interatomic forces depend linearly on relative displacements would have no coefficient of expansion. A derivation based on standard results of thermodynamics leads to the result (Becker (ref. 6.1))

$$C_p - C_V = TV\beta^2/\kappa \tag{6.29}$$

where β is the volume coefficient of expansion and κ is the compressibility. The difference is negligible at low temperatures, but becomes appreciable for temperatures comparable with θ_D. For example for lead at room temperature, $C_p = 6{\cdot}40$ and $C_V = 6{\cdot}03$ cal . deg^{-1} mole^{-1}. This is an indication that anharmonic effects cannot be neglected in attempting to predict the specific heat over a wide temperature range. The theory is too elaborate to give here. Müller and

Brockhouse (ref. 6.5) made neutron measurements on copper, and were able to allow for anharmonicity in calculating C_p by measuring the change of the frequencies $\omega_j(\mathbf{q})$ with temperature. The dotted line in Fig. 6.8 gives C_p predicted on the basis of the harmonic approximation while the full line includes an allowance for anharmonic interactions. The points, which show the results of direct measurement of C_p, lie above the line by an amount which agrees well with the expected electronic contribution C_e. The processes

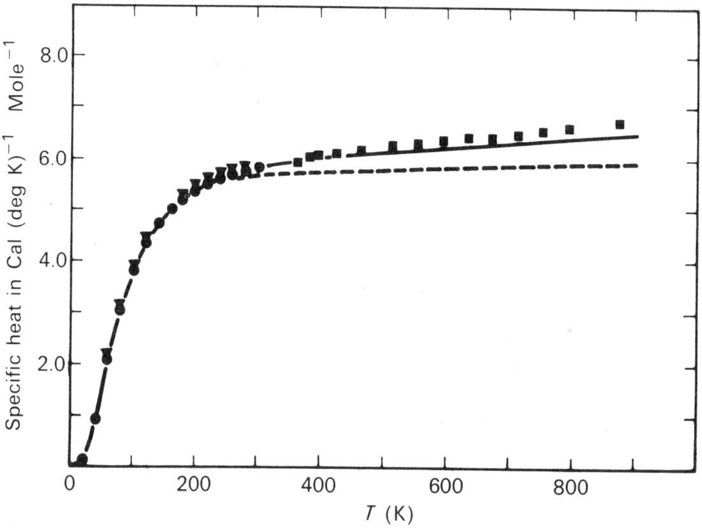

Figure 6.8 The points show experimental values of the specific heat C_p of copper. The dotted line gives the results of theory using the harmonic approximation; the full line includes a correction for anharmonicity. (After Müller and Brockhouse, 1971, *Canad. J. Phys.* 49, 704.)

contributing to the specific heat of copper over a temperature range of 800 K are therefore apparently well understood.

6.4 Thermal conductivity

As we have just seen, the specific heat of a crystal is fairly well accounted for without going beyond the harmonic approximation, and the theory is relatively simple because only phonon frequencies are involved. Thermal conductivity on the other hand cannot be understood at all without taking into account the interaction of the lattice waves with the boundaries of the crystal, with impurity atoms and defects such as dislocations, and with one another. In the absence of such interactions, thermal energy travels through a solid with the group velocity of the waves, but in practice the flow of heat is usually a diffusive process. The thermal conductivity σ_t is defined by

$$Q = \sigma_t \frac{\partial T}{\partial x} \tag{6.30}$$

where Q is the thermal energy crossing a unit area which is perpendicular to the x direction, per unit time. With almost as many factors involved as determine the weather, it is not surprising that there is no 'exact' theory of thermal conductivity, but only approximate theories which apply in particular ranges of temperature. In a metal the conduction electrons are generally more effective in transporting energy than are the phonons and the electron-phonon interaction provides yet another source of phonon scattering; at this stage however our discussion will be confined to insulators.

The heat current in the x direction is the sum of the energies carried by all the modes

$$Q = \sum_{\mathbf{q}j} n_j(\mathbf{q})\hbar\omega_j(\mathbf{q})(u_j(\mathbf{q}))_x \tag{6.31}$$

considering a unit volume of the crystal. Here $n_j(\mathbf{q})$ is the phonon occupation number which we met in Chapter 3, its average value being

$$\bar{n}_j(\mathbf{q}) = \left[\exp\left(\frac{\hbar\omega_j(\mathbf{q})}{k_B T}\right) - 1\right]^{-1} \tag{6.32}$$

while $u_j(\mathbf{q})$ is the group velocity, the gradient of the dispersion curve. Clearly under equilibrium conditions, with a uniform temperature, $Q = 0$ since the group velocities are distributed isotropically. However in the absence of processes acting to restore equilibrium, a current once established would persist. As an approximation, it may be assumed that $n_j(\mathbf{q})$ relaxes to its equilibrium value according to the equation (ref. 6.6),

$$\left[\frac{\partial n_j(\mathbf{q})}{\partial t}\right] = \frac{\bar{n}_j(\mathbf{q}) - n_j(\mathbf{q})}{\tau_j(\mathbf{q})} \tag{6.33}$$

where $\tau_j(\mathbf{q})$ is a relaxation time. The square brackets denote that the time rate of change is due to interactions only. The transport equation may now be taken to be

$$\left[\frac{\partial n_j(\mathbf{q})}{\partial t}\right] - (u_j(\mathbf{q}))_x \frac{\partial T}{\partial x}\left(\frac{dn_j(\mathbf{q})}{dT}\right) = 0 \tag{6.34}$$

This expresses the condition that as a wave packet moves, the occupation number must change with time through collision processes to be appropriate to the temperature of the region in which it finds itself. In the second term, $n_j(\mathbf{q})$ can be approximated by $\bar{n}_j(\mathbf{q})$. Combining equations 6.33 and 6.34 we then obtain

$$n_j(\mathbf{q}) - \bar{n}_j(\mathbf{q}) = -\tau_j(\mathbf{q})(u_j(\mathbf{q}))_x \frac{\partial T}{\partial x}\left(\frac{d\bar{n}_j(\mathbf{q})}{dT}\right) \tag{6.35}$$

Substituting this result into Equ. 6.31 and noting that the equilibrium value $\bar{n}_j(\mathbf{q})$ gives no heat flow, we obtain for the heat flow in the (positive) x direction

$$Q = -\sum_{\mathbf{q}j} \hbar\omega_j(\mathbf{q})\tau_j(\mathbf{q})(u_j(\mathbf{q}))_x^2 \left(\frac{d\bar{n}_j(\mathbf{q})}{dT}\right)\left(\frac{\partial T}{\partial x}\right) \tag{6.36}$$

For an isotropic material $(u_j(\mathbf{q}))_x^2$ may on the average be replaced by $\frac{1}{3}u_j^2(\mathbf{q})$, so that on comparing equations 6.36 and 6.30,

$$\sigma_t = \frac{1}{3}\sum_{\mathbf{q}j} \hbar\omega_j(\mathbf{q})\tau_j(\mathbf{q})u_j^2(\mathbf{q})\left(\frac{d\bar{n}_j(\mathbf{q})}{dT}\right) \tag{6.37}$$

However $\hbar\omega_j(\mathbf{q})(d\bar{n}_j(\mathbf{q})/dT)$ is just the change in energy with temperature of one mode (see Equ. 6.2), i.e. its contribution to the specific heat, which we write as $C_j(\mathbf{q})$. Thus Equ. 6.37 becomes

$$\sigma_t = \tfrac{1}{3}\sum_{\mathbf{q}j} \tau_j(\mathbf{q})u_j^2(\mathbf{q})C_j(\mathbf{q}) \tag{6.38}$$

If therefore we were further to assume that the group velocity and relaxation time are the same for all modes (which clearly is generally a poor approximation) Equ. 6.38 would reduce to

$$\sigma_t = \tfrac{1}{3}\tau u^2 C \tag{6.39}$$

where C is the specific heat (of unit volume), $\sum_{\mathbf{q}j} C_j(\mathbf{q})$. This equation can also be written

$$\sigma_t = \tfrac{1}{3}\lambda u C \tag{6.40}$$

where $\lambda = \tau u$ is a mean free path between collisions in which one phonon of energy is exchanged. The same equation gives the thermal conductivity of an ideal gas, and is often used as the starting point for a discussion of the thermal conductivity of a solid. Its use tends to obscure the fact that phonons of different properties contribute to σ_t in different ways, and to very different extents.

6.5 Relaxation times in insulators

In Equ. 6.38 the only quantity which is not known when the dispersion relation $\omega_j(\mathbf{q})$ is known, is $\tau_j(\mathbf{q})$. In detailed theories of the interaction of phonons with imperfections and with the boundaries, the Debye approximation is usually introduced, that is the dispersion curves are approximated by a single acoustic branch, specified by a constant velocity u (Equ. 6.15) which is an appropriate average over the transverse and longitudinal branches. When two distinct scattering processes operate simultaneously, having relaxation times τ_1 and τ_2 for a particular mode, it is assumed that the scattering rates are additive, so that

$$\tau^{-1} = \tau_1^{-1} + \tau_2^{-1} \tag{6.41}$$

When one process has a much shorter relaxation time than others it is therefore the only one which need be considered.

At very low temperatures, only modes of low frequency are thermally excited and these are scattered only by the boundaries. The mean free path is the same for all, approximately the diameter of a cylindrical specimen. We then expect Equ. 6.40 to apply. Fig. 6.9 (Berman *et al.* (ref. 6.7)) shows σ_t measured over a range of temperature for Al_2O_3. The three curves correspond to three specimens of different diameters prepared from the same single crystal. Al_2O_3 has a high Debye temperature, about 1000 K, and the T^3 law for C is valid below 30 K. For the purest crystals σ_t was indeed found to vary as T^3 below 10 K, and λ determined from Equ. 6.40 was approximately equal to the diameter of the specimen in each case. When a large crystal held at a low temperature (1 or 2 K) has a resistive layer of relatively small area deposited on a plane face, and an equally small layer of superconducting material on an opposite face to serve as a bolometer, a pulse of current through the resistance results in two heat pulses being detected at the bolometer. These are wave packets which have travelled 'ballistically' through the crystal with the group velocities of transverse and longitudinal modes respectively.

As the temperature increases, modes having a larger value of q acquire an appreciable value of $C_j(\mathbf{q})$ and begin to contribute to σ_t. Such modes are scattered by defects in the crystal; σ_t therefore increases more slowly with T and eventually reaches a maximum value which depends strongly on the impurity and defect content of the crystal. This is apparent from Fig. 6.10 which shows results obtained by Jackson and Walker (ref. 6.8) for very pure crystals of NaF. This material was chosen because both Na and F occur

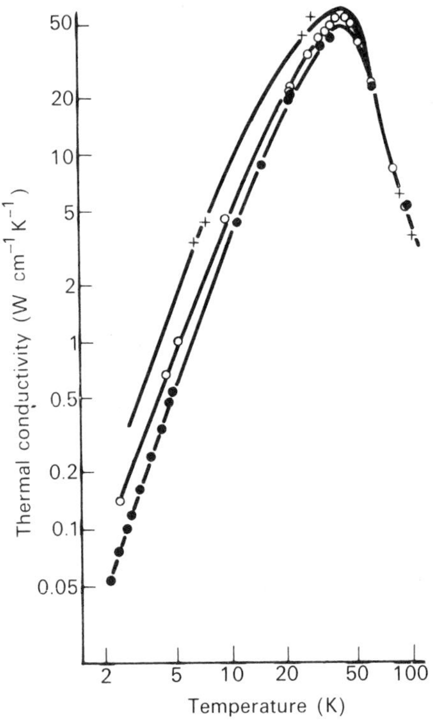

Figure 6.9 The thermal conductivities of cylindrical specimens of Al_2O_3 having diameters 1·02, 1·55 and 2·8 mm respectively. ● 1·02 mm, ○ 1·55 mm, + 2·8 mm. (After Berman, Foster and Ziman, 1955, *Proc. Roy. Soc. A* **231**, 130.)

naturally as single isotopes; the random variation of isotopic mass can be quite an important scattering mechanism which thus need not be considered for NaF. The material giving curve A had only about one part per million of impurities, mainly Ca and K, and curve B, 6 p.p.m. From the detailed analysis, which involved estimates of relaxation times for impurity scattering and dislocation scattering, and their dependence on phonon frequency, Jackson and Walker concluded that if all impurities could be eliminated the peak value of σ_t would be 310 watt cm^{-1} deg^{-1}, while if dislocations were absent it would rise further to 400 watt cm^{-1} deg^{-1}!

Beyond about 30 K in Fig. 6.10 the curves for crystals of different impurity content run together again, showing that another scattering mechanism is dominant. This is the mutual scattering of the lattice waves, which we recall does not occur in the harmonic approximation. The theory, first developed in detail by Peierls (ref. 6.9) is complicated, and only a qualitative account will be

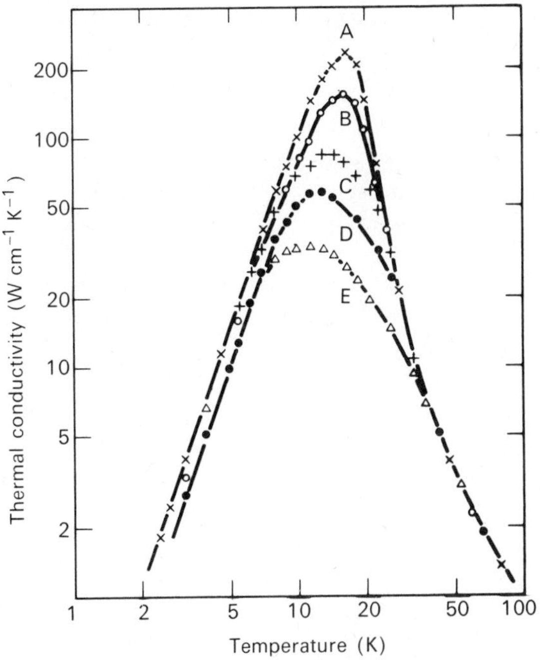

Figure 6.10 The thermal conductivities of five specimens of NaF having different impurity contents. (After Jackson and Walker, 1971, *Phys. Rev. B* **3**, 1428.)

given. The most important process is one in which two wave packets in passing through the same region contribute energy to a third. In particle terms, two phonons collide to create a third phonon, with conservation of energy. The converse process in which a quantum of energy is lost from one mode and one is gained in each of two others is equally important. Besides the energy condition

$$\omega_{j1}(\mathbf{q}) + \omega_{j2}(\mathbf{q}_2) = \omega_{j3}(\mathbf{q}_3) \tag{6.42}$$

it is found that

$$\mathbf{q}_1 + \mathbf{q}_2 = \mathbf{q}_3' \tag{6.43}$$

must be satisfied, often referred to as 'conservation of pseudo-momentum' (§3.2). The wavevector of the phonon produced is written as $\mathbf{q}_3'$ since it may lie outside the first Brillouin zone. It is then equivalent to a phonon of wavevector $\mathbf{q}_3$ obtained by subtracting a vector $\mathbf{K}_h$ of the reciprocal lattice. We therefore rewrite Equ. 6.43 as

$$\mathbf{q}_1 + \mathbf{q}_2 = \mathbf{q}_3 + \mathbf{K}_h \tag{6.44}$$

with all three wavevectors inside the zone. It is not generally possible to satisfy equations 6.42 and 6.44 simultaneously with phonons belonging to the same branch. An interaction for which $\mathbf{q}_3 = \mathbf{q}_3'$, i.e. $K_h = 0$, is referred to as a normal

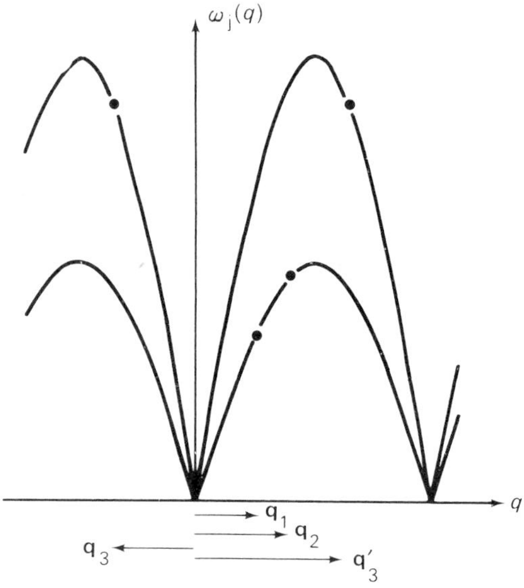

Figure 6.11 Illustrates an umklapp process. The group velocity (slope of the dispersion curve) is positive at q_1 and at q_2, but is negative at q_3' or q_3.

or N-process. In such a process the product of frequency and group velocity is approximately unchanged and therefore (Equ. 6.31) the heat flow is unaffected. However, such processes do contribute to maintaining the 'equilibrium' phonon distribution that corresponds to a steady heat flow, and their relaxation times appear in the detailed theory. 'Umklapp' or U-processes for which $K_h \neq 0$ are however more effective. Such a process is illustrated in Fig. 6.11, where for convenience all the wavevectors are taken to be in the same direction. The group velocity $\mathbf{u}_3$ is in the *opposite* direction to $\mathbf{u}_1$ and $\mathbf{u}_2$. Such processes are therefore very important in determining σ_t. If K_D is the 'radius' of the Brillouin zone, $\mathbf{q}_1$ and $\mathbf{q}_2$ must be at least $\frac{1}{2}K_D$ for a U-process, and their energies about $\frac{1}{2}k_B\theta_D$, using the Debye approximation. The occupation number for such phonons is

$$\left\{\exp\left(\frac{\theta_D}{2T}\right) - 1\right\}^{-1} \triangleq \exp\left(-\frac{\theta_D}{2T}\right) \tag{6.45}$$

We might expect the relaxation time for U-processes to be inversely proportional to the population of phonons capable of bringing about such processes. This is found to be the case; the experimental variation of the relaxation time is so great compared with the variation with temperature of the phonon specific heat that σ_t itself shows a temperature dependence

$$\sigma_t \propto \exp(\alpha\theta_D/T) \tag{6.46}$$

where α is less than one. For the purest crystal of NaF (curve A of Fig. 6.10) Equ. 6.46 was found to apply accurately between 20 K and 35 K, and this is found to be the case for many materials in the region beyond the peak in the thermal conductivity.

At temperatures comparable with the Debye temperature, $C_j(\mathbf{q})$ tends to the constant value k_B for all phonons, and the occupation number becomes proportional to T. Thus $\tau_j(\mathbf{q})$ and σ_t become inversely proportional to T, in agreement with experiment.

6.6 Other factors which influence σ_t

We have already mentioned that the presence of different isotopes of an element results in an additional scattering mechanism. Fig. 6.12 shows how the conductivity of isotopically enriched Ge^{74} substantially exceeds that of ordinary Ge, (ref. 6.10). The thermal conductivity of alloy systems, such as Ge–Si, is less than that of either of the pure materials. Grossly disordered solids such as glasses have low thermal conductivities, one or two orders of magnitude less than those of single crystals of typical ionic materials at room temperature. Values for the latter, incidentally, are often in the range 0·5 to 0·1 watt cm^{-1} deg^{-1}, very different from the peak values which we saw in §6.5.

The topic of thermal conductivity in metals lies somewhat outside the scope of this book and we therefore make only a few comments. The presence of conduction electrons gives an additional mechanism for phonon scattering. It turns out that at ordinary temperatures the relaxation time for this process is long compared with that for umklapp scattering by phonons, so that the influence of the electrons on the lattice contribution to σ_t is not important. However at temperatures below the maximum in the thermal conductivity, the phonon relaxation time is substantially decreased by electron scattering. We may therefore conclude that the lattice contribution to σ_t is less, and at low temperatures considerably less, than it would be in an insulator of similar dynamical properties.

The contribution of the electrons to the thermal conductivity is determined by an equation similar to Equ. 6.39; the quantities on the right now of course refer to electrons. As we saw in §6.3, the electron specific heat is small compared with the lattice specific heat except at low temperatures. However the velocity

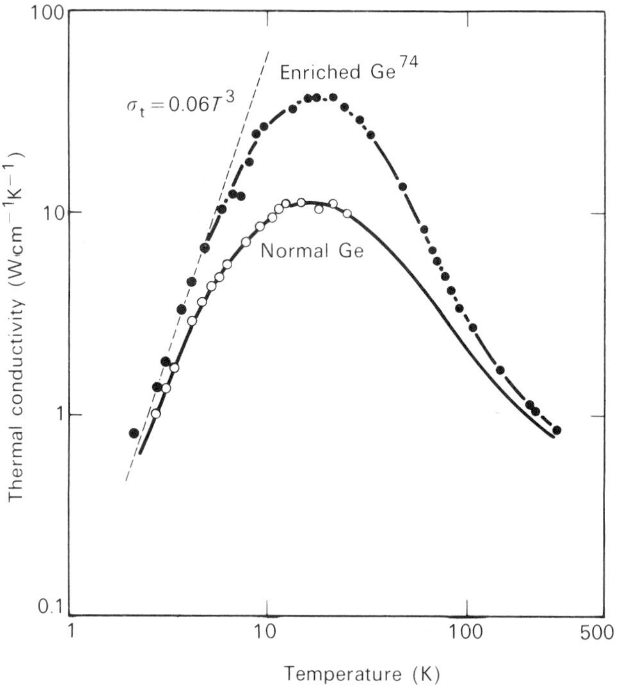

Figure 6.12 The thermal conductivity of germanium is appreciably increased when it is composed mainly of only one isotope. (After Geballe and Hull, 1958, *Phys. Rev.* **110**, 773.)

of the electrons is two or three orders of magnitude greater than phonon velocities since only those electrons with wavevectors near the Fermi surface contribute. The relaxation time for electron collisions is determined by interaction with the boundaries, with impurities and imperfections, with phonons and with other electrons. Detailed calculations give the result that for pure metals the electron contribution to σ_t is generally at least an order of magnitude greater than the phonon contribution over the whole temperature range, so that the latter contribution can be neglected. In this situation both charge and heat currents are carried by the electrons only. Since similar factors are involved in both transport processes, the ratio of thermal conductivity σ_t to electrical conductivity σ_e is the same at a fixed temperature for many metals,

$$\frac{\sigma_t}{\sigma_e} = \frac{\pi^2 k_B^2 T}{3e^2} \tag{6.47}$$

This is the well-known Wiedemann–Franz law.

References and suggestions for further reading

6.1 Becker, R. 1967, *Theory of Heat*, Springer, Berlin.
6.2 Blackman, M. 1941, *Rep. Prog. Phys.* **8**, 11.
6.3 Mott, N. F. and Jones, H. 1936, *The Theory of the Properties of Metals and Alloys*, O.U.P., Oxford
6.4 Cowley, R. A., Woods, A. D. B., and Dolling, G. 1966, *Phys. Rev.* **150**, 487.
6.5 Müller, A. P. and Brockhouse, B. N. 1971, *Can. J. Phys.* **49**, 704.
6.6 Klemens, P. G. 1958, *Solid State Physics* 7, 1, Academic Press, New York.
6.7 Berman, R., Foster, E. L., and Ziman, J. M. 1955, *Proc. R. Soc.* **A231**, 130.
6.8 Jackson, H. E. and Walker, C. T. 1971, *Phys. Rev.* **B3**, 1428.
6.9 Peierls, R. E. 1956, *Quantum Theory of Solids*, Clarendon Press, Oxford.
6.10 Geballe, T. H. and Hull, G. W. 1958, *Phys. Rev.* **110**, 773.

7

Dielectric and Optical Properties

7.1 Dielectric constant and refractive index

In the M.K.S. system of units the force between point charges in a vacuum is

$$F = \frac{Q^2}{4\pi\epsilon_0 R^2} \tag{7.1}$$

where $(1/4\pi\epsilon_0) = 9 . 10^9$ N m^2 coulomb^{-2}. The relation between polarization and electric field in an isotropic medium is

$$P = \epsilon_0(\epsilon - 1)E \tag{7.2}$$

where ϵ is the dielectric constant. It is important to note that E is the field in the medium, which we can define in a crystal by averaging over a distance large compared with the unit cell dimension, to eliminate variations on an atomic scale, and that this field may differ from the applied field E_A. For example, consider the well defined (if impractical) situation illustrated in Fig. 7.1a where a uniform field E_A is established midway between two widely separated fixed point charges. If a cylindrical specimen of small diameter is placed in the field as in Fig. 7.1b the field inside the specimen is E_A. If however a disc shaped specimen is placed as in Fig 7.1c the surface charges of density $\pm P$ produce a field P/ϵ_0 inside the specimen (as can readily be confirmed by integration over the surfaces) which opposes the applied field. Hence from Equ. 7.2,

$$P = \epsilon_0(\epsilon - 1)\left(E_A - \frac{P}{\epsilon_0}\right), \text{ that is}$$

$$P = \epsilon_0\left(1 - \frac{1}{\epsilon}\right)E_A \tag{7.3}$$

in this instance. We may contrast this situation with that where a uniform applied field is established between the plates of a capacitor connected to a battery. The field in a vacuum is (V/d) where V is the potential difference and d the distance

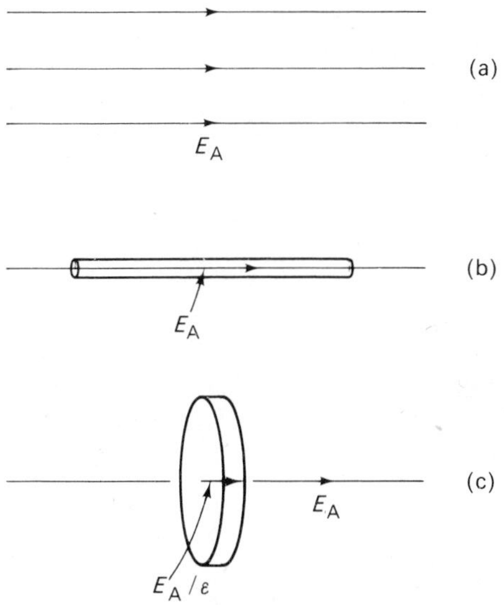

Figure 7.1 Needle shaped and disc shaped specimens in a uniform applied field E_A. The fields inside the specimens are respectively E_A and $E_A - P/\epsilon_0 = E_A/\epsilon$.

between the plates. When a disc shaped specimen of thickness d is inserted between the plates, the field inside the specimen remains V/d because the effect of the surface polarization charges is exactly cancelled by additional charges supplied by the battery to the capacitor plates. In what follows we assume that the situation is such that surface charges do not affect the field. When the field E varies as $\exp(-i\omega t)$ the polarization P will not necessarily vary in phase. We take account of this by writing

$$\epsilon(\omega) = \epsilon'(\omega) + i\epsilon''(\omega) \tag{7.4}$$

Maxwell's equations lead to a wave equation for the electric field in a material medium which is

$$\frac{\partial^2 E}{\partial x^2} = \mu\mu_0\epsilon\epsilon_0 \frac{\partial^2 E}{\partial t^2} \tag{7.5}$$

In a vacuum the velocity of an electromagnetic wave is thus given by

$$c = (\epsilon_0\mu_0)^{-1/2} \tag{7.6}$$

Setting the magnetic permeability $\mu = 1$, an excellent approximation for most materials, the velocity in the material is

$$c' = (\epsilon\epsilon_0\mu_0)^{-1/2} = \frac{c}{n} \tag{7.7}$$

Hence we have the well known relation between the refractive index n and dielectric constant ϵ that

$$n^2 = \epsilon \tag{7.8}$$

It should be noted that these quantities can only be equated at the same frequency; contrast the static dielectric constant of water, which is 80, with the square of the index of refraction at optical frequencies, which is 1·7. Writing

$$n = n' + in'' \tag{7.9}$$

we see that

$$(n')^2 - (n'')^2 = (\epsilon')^2 \quad \text{and} \quad 2n'n'' = \epsilon'' \tag{7.10}$$

The variation of the electric field in the wave may thus be written

$$\exp\left\{-i\omega\left(t - \frac{nx}{c}\right)\right\} = \exp\left\{-i\omega\left(t - \frac{n'x}{c}\right)\right\}\exp\left(\frac{-\omega n''x}{c}\right) \tag{7.11}$$

from which we see that n'' determines the absorption coefficient.

7.2 Electronic polarizability and optical dielectric constant

The dielectric constant of ionic crystals may be divided into two parts. One of these depends entirely on the response of the electrons to a field. We shall denote it $\epsilon(\infty)$ since it remains almost constant up to frequencies in the optical region of the electromagnetic spectrum, $(\omega/2\pi) \sim 10^{15}$ sec^{-1}, much higher than frequencies in the phonon region with which we are primarily concerned. In principle $\epsilon(\infty)$ can be calculated if the electron energy levels and wave functions are known, but a simple model is reasonably satisfactory for materials with widely spaced energy bands, when the electrons can be regarded as well localized on atomic sites. This model consists in ascribing an electronic polarizability α to each atom or ion, its dipole moment being given by

$$p = \epsilon_0 \alpha E_l \tag{7.12}$$

where E_l is the field acting on the atom at site l, which will generally differ from the field E.

The second contribution to $\epsilon(\omega)$ comes from relative displacement of positive and negative ions, and falls to unity at frequencies above those of the lattice vibrations ($\omega/2\pi \sim 5 \,.\, 10^{12}$ sec^{-1}), where the ions are 'clamped by their own inertia' in the varying field. We therefore write

$$\epsilon(\omega) - 1 = (\epsilon_{\rm I}(\omega) - 1) + (\epsilon(\infty) - 1) \tag{7.13}$$

to take account of the polarization arising from both mechanisms. In a crystal such as germanium or argon, clearly $\epsilon(\omega) = \epsilon(\infty)$ from $\omega = 0$ through the range

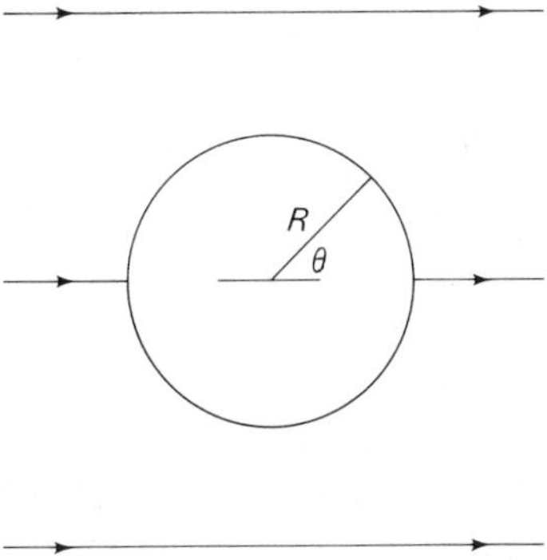

Figure 7.2 Spherical cavity in a uniformly polarized medium. The surface charge density at (R, θ) is $P \cos \theta$.

of phonon frequencies. (It is a little confusing that $\epsilon_{\rm I}(\omega) = 1$ corresponds to *no* ionic contribution to the polarization, but see Equ. 7.2.)

For the remainder of this section we return to a discussion of $\epsilon(\infty)$. Consider a crystal of, for example, argon, which has the f.c.c. structure with one atom per primitive unit cell of volume v. In a uniform applied field E each atom will acquire a dipole moment given by Equ. 7.12, and therefore

$$P = \epsilon_0 \alpha E_l / v \tag{7.14}$$

Draw a sphere of radius R about the atom at site l, where R is large compared with the unit cell dimension. The material outside the sphere, being at a considerable distance, can be regarded as continuously polarized and therefore acts through the surface charge density $P \cos \theta$ produced on the sphere (see Fig. 7.2). Integration over this surface gives the field produced at the centre of the sphere as

$$E_{\rm L} = \int_0^{\pi} \frac{2\pi R \sin \theta R d\theta P \cos \theta \cos \theta}{4\pi\epsilon_0 R^2} = \frac{P}{3\epsilon_0} \tag{7.15}$$

This is called the Lorentz field. The symmetry of the arrangement of dipoles *inside* the sphere is such that they produce no field at the centre.

$$\therefore E_l = E + \frac{P}{3\epsilon_0} \tag{7.16}$$

The cancellation, at the centre of the sphere, of the fields of the dipoles, is valid only for certain crystal structures, but is should apply in a liquid or a gas where the distribution inside the sphere is uniform when averaged over time. Combining equations 7.16, 7.14 and 7.2 we obtain

$$\epsilon(\infty) - 1 = \frac{\alpha/v}{1 - (\alpha/3v)} \tag{7.17}$$

that is

$$\frac{\epsilon(\infty) - 1}{\epsilon(\infty) + 2} = \frac{\alpha}{3v} \tag{7.18}$$

This is the Clausius-Mosotti formula. For argon of course, $\epsilon(\infty)$ may equally well be replaced by $\epsilon(0)$. Fig. 7.3 shows that for argon (ref. 7.2) the experimental values of $\dfrac{\epsilon(0) - 1}{(\epsilon(0) + 2)D}$ where D is the density, do not vary by more than 2% over a wide range of density (i.e. of v) covering the gaseous, liquid and solid phases, in good agreement with the assumptions that α is constant and the Lorentz field is $P/3\epsilon_0$.

The same value of the Lorentz field should apply in an alkali halide, provided of course that the assumption of nonoverlapping electron distributions is valid. If α_1 and α_2 are the polarizabilities of positive and negative ions respectively, the Clausius-Mosotti formula becomes

$$\frac{\epsilon(\infty) - 1}{\epsilon(\infty) + 2} = \frac{\alpha_1 + \alpha_2}{3v} \tag{7.19}$$

Tessman *et al.* (ref. 7.1) showed that it was possible to choose constant values of α for each of four negative ions (F, Cl, Br, I) and five positive ions (Li, Na, K, Rb, Cs) so as to reproduce the measured values of $\dfrac{3v(\epsilon(\infty) - 1)}{\epsilon(\infty) + 2}$ for twenty alkali halides (LiF, . . ., CsI) with a mean deviation of no more than 1%, except for the fluorides where the mean deviation was about 8%.

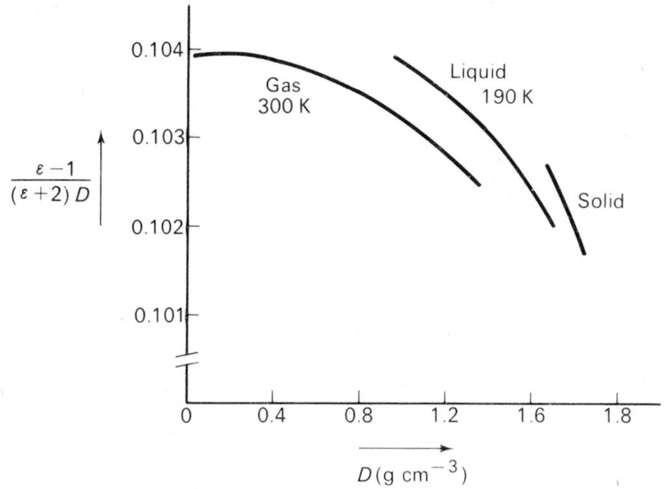

Figure 7.3 The quantity $(\epsilon - 1)/(\epsilon + 2)D$ varies by less than 2% over a wide range of density for argon. (From data provided by W. B. Daniels.)

7.3 Ionic contribution to the dielectric constant

We introduce this topic by considering a model for an ionic crystal which, although it has obvious shortcomings, does enable some important points to be brought out quite simply. This is the rigid ion model, already mentioned in §5.4, and which we saw there gave quite a passable explanation of the general shape of the phonon dispersion curves for KBr (dotted lines in Fig. 4.4) as well as giving good agreement with the observed cohesive energies of alkali halides. Since the ions are treated as point charges (which we take to be $\pm Ze$) there is no electronic polarizability and $\epsilon(\infty) = 1$. This is the major deficiency of the model.

Displacing a positive ion through a distance $\mathbf{u}$ must produce the same effect on the field everywhere in the crystal as keeping the ion fixed, with the addition of a charge $-Ze$ in the original position and $+Ze$ in the displaced position. In other words the displacement is equivalent to the creation of a dipole $Ze\mathbf{u}$. When all positive and all negative ions are displaced by amounts u_1 and u_2 (in the same direction) the polarization is therefore

$$P = Ze(u_1 - u_2)/v \tag{7.20}$$

Let $\beta\mathbf{u}$ be the restoring force on an ion when it is displaced a distance $\mathbf{u}$ relative to its nearest neighbours, originating in the short range repulsive potential. β can be expressed in terms of A and ρ of Equ. 5.17, but we leave it simply as a parameter.

In an applied field E the equation of motion of a positive ion will be

$$m_1 \frac{\partial^2 u_1}{\partial t^2} = Ze\left(E + \frac{P}{3\epsilon_0}\right) - \beta(u_1 - u_2)$$

and of the negative ion

$$m_2 \frac{\partial^2 u_2}{\partial t^2} = -Ze\left(E + \frac{P}{3\epsilon_0}\right) - \beta(u_2 - u_1) \tag{7.21}$$

Introducing a reduced mass

$$m = \frac{m_1 m_2}{m_1 + m_2} \tag{7.22}$$

and using Equ. 7.20 we obtain

$$m \frac{\partial^2 P}{\partial t^2} = \frac{(Ze)^2}{v}\left(E + \frac{P}{3\epsilon_0}\right) - \beta P \tag{7.23}$$

If E is now assumed to vary as $\exp(-i\omega t)$, P must vary in phase, and Equ. 7.23 gives

$$\frac{P}{E} = \epsilon_0(\epsilon(\omega) - 1) = \frac{(Ze)^2/v}{\beta - \dfrac{(Ze)^2}{3v\epsilon_0} - m\omega^2} \tag{7.24}$$

The dielectric constant $\epsilon(\omega)$ given by this equation is sketched in Fig. 7.4. It diverges at a frequency ω_T given from Equ. 7.24 by

$$m\omega_T^2 = \beta - \frac{(Ze)^2}{3v\epsilon_0} \tag{7.25}$$

Evidently at this frequency the crystal is resonating by being driven in one of the normal modes of vibration. We also see this by putting $E = 0$ in Equ. 7.23, when the solution for the frequency of free oscillation is Equ. 7.25. The mode

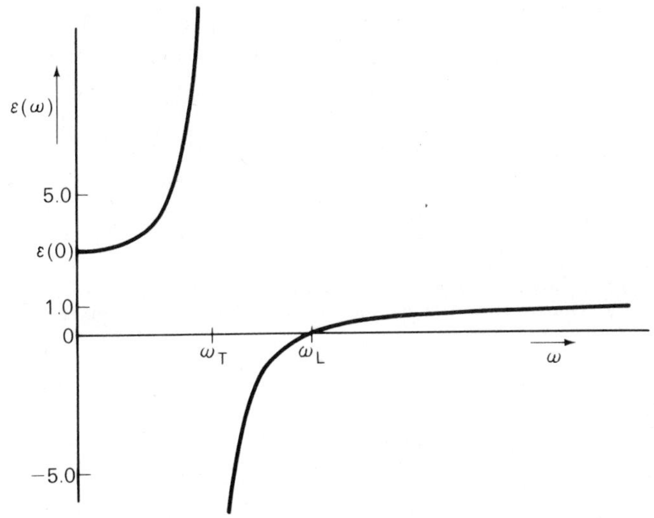

Figure 7.4 The dielectric constant as a function of frequency for the rigid ion model of an ionic crystal, for which $\epsilon(\infty) = 1$. $\epsilon(0)$ has been given the value 3.00.

involved must be an optic mode since the two types of atom are moving against one another, (§3.5 and Fig. 3.7). Furthermore, Equ. 7.23 assumes there is no 'depolarizing' field $-P/\epsilon_0$, as discussed in §7.1. For the transverse mode shown in Fig. 7.5 it is fairly clear that there is no depolarizing field since the situation is analogous to that of thin sheets alternately polarized parallel to a direction lying in their surfaces. Since however we have taken P to be spatially uniform in Equ. 7.23 we must take the wavelength $2\pi/q$ in Fig. 7.5 to be large compared with the unit cell dimension, while remaining small compared with any dimension of the crystal. We have therefore identified the mode of frequency ω_T as the transverse optic mode for which $q \to 0$. For a general direction of $\mathbf{q}$ there are two optic branches in an alkali halide which approach the same frequency as $q \to 0$. This is shown schematically in Fig. 7.7.

Consider now a longitudinal optic mode for which $2\pi/q \gg a$, with polarization varying cosinusoidally as indicated in Fig. 7.6. It can be shown that there is, in the absence of an applied field, a cosinusoidally varying field given by $E = -P/\epsilon_0$. This is plausible since the situation is analogous to a stack of thin sheets alternately polarized perpendicular to their surfaces. The equation for the polarization, with $q \to 0$, is

$$m\frac{\partial^2 P}{\partial t^2} = \frac{(Ze)^2}{v}\left(-\frac{P}{\epsilon_0} + \frac{P}{3\epsilon_0}\right) - \beta P \tag{7.26}$$

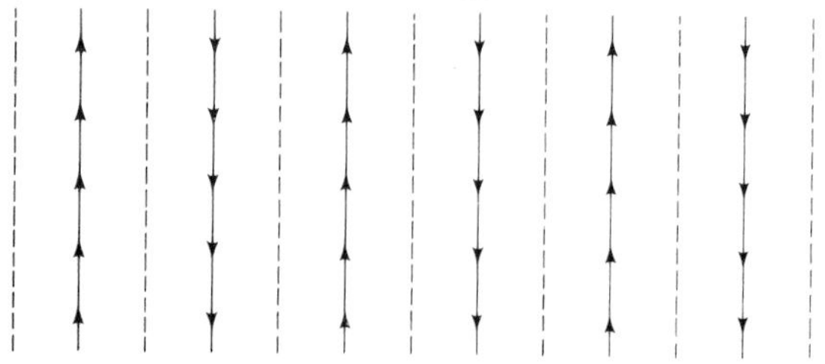

Figure 7.5 Nodes (dotted lines) and antinodes (solid lines) of polarization in a transverse wave of long wavelength. Arrows denote the direction of polarization; the wavevector **q** is horizontal.

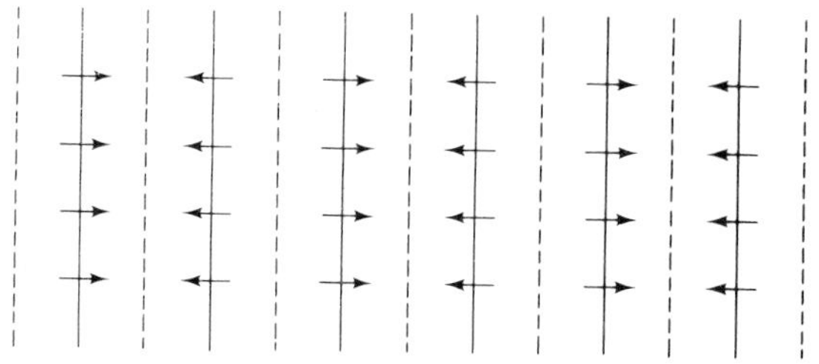

Figure 7.6 As for Fig. 7.5, but for a longitudinal wave.

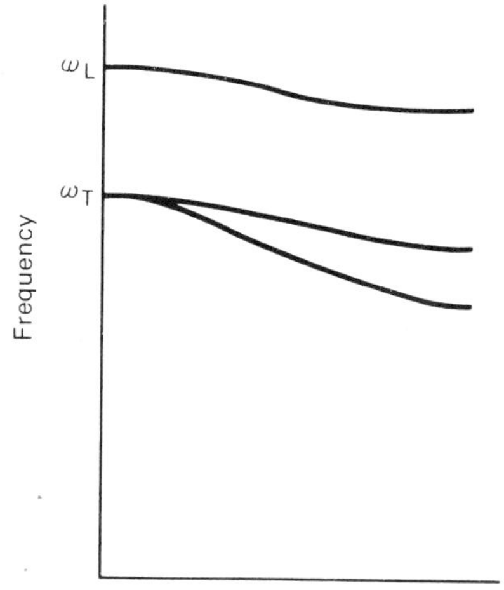

Figure 7.7 In a diatomic cubic crystal the two transverse optic branches approach the same frequency ω_T as $q \to 0$.

and the frequency of the mode is given by

$$m\omega_L^2 = \beta + \frac{2(Ze)^2}{3v\epsilon_0} \tag{7.27}$$

It follows that $\omega_L > \omega_T$. If we substitute this frequency in Equ. 7.24 we obtain

$$\epsilon(\omega_L) = 0$$

This can also be seen from Equ. 7.2; $E = -P/\epsilon_0$ requires $\epsilon = 0$. A mode with these characteristics occurs also for electrons in a metal and for an ionized gas, when it is referred to as a plasma mode. Note that for $Z = 0$ the longitudinal and transverse optic branches are predicted to have the same frequency for $q \to 0$, as is the case in silicon for example (Fig. 5.5). The 'splitting' of these branches is a characteristic of Coulomb interaction.

On combining equations 7.24 and 7.25 and putting $\omega = 0$ we have a relation between measurable quantities

$$\epsilon(0) - 1 = \frac{(Ze)^2}{v\epsilon_0 m\omega_T^2} \tag{7.28}$$

For KBr with $Z = 1$ the two sides are respectively 3·46 and 1·86. We may suspect that the discrepancy is caused by our neglect of the electronic polarizability.

7.4 'Realistic' model for an alkali halide

It is not difficult to combine the models described in §7.2 and 7.3. That is, the ions are taken to have charges $\pm Ze$ and polarizabilities α_1 and α_2 respectively. We shall however not give derivations but simply quote the results, which are

$$\epsilon(\omega) = \epsilon(\infty) + \frac{(\epsilon(\infty) + 2)^2 (Ze)^2}{9v\epsilon_0 m(\omega_T^2 - \omega^2)} \tag{7.29}$$

$$m\omega_T^2 = \beta - \frac{(\epsilon(\infty) + 2)(Ze)^2}{9v\epsilon_0} \tag{7.30}$$

$$m\omega_L^2 = \beta + \frac{2(\epsilon(\infty) + 2)(Ze)^2}{9v\epsilon_0 \epsilon(\infty)} \tag{7.31}$$

while $\epsilon(\infty)$ is given by Equ 7.19. These equations satisfy a number of expected conditions such as $\epsilon(\omega_T) = \infty$, $\epsilon(\omega_L) = 0$ and $\epsilon(\omega) \to \epsilon(\infty)$ for $\omega \gg \omega_T$. On putting $\epsilon(\infty) = 1$ they reduce to those which we had in §7.3.

It is readily shown from equations 7.29, 7.30 and 7.31 that

$$\frac{\omega_L^2}{\omega_T^2} = \frac{\epsilon(0)}{\epsilon(\infty)} \tag{7.32}$$

This particular result, first obtained by Lyddane, Sachs and Teller, does not however depend on the model used but is a phenomenological relation. It is well satisfied by the experimental results; the two ratios for KBr, for example, are 1·93 and 1·90 respectively. From Equ. 7.29 we obtain a result which does depend on the model,

$$\epsilon(0) - \epsilon(\infty) = \frac{(\epsilon(\infty) + 2)^2 (Ze)^2}{9v\epsilon_0 m\omega_T^2} \tag{7.33}$$

For KBr at 90 K the experimental values of the quantities on the left and right respectively are 2·10 and 3·92. This is a serious discrepancy because we have apparently not omitted anything from consideration that should be important. To satisfy Equ. 7.33 it is necessary to reduce Z to an 'effective' value of 0·73, in apparent disagreement with various other items of experimental evidence which suggest that alkali halides are fully ionic, or almost so.

This model has to be abandoned not only because of this discrepancy, but because it completely fails to predict the measured values of $\omega_j(\mathbf{q})$, giving much worse agreement with the measured values than does the rigid ion model. As already mentioned in §5.4, it must be assumed that the electronic dipole moment of an ion depends not only on the field E_l but also on the positions of nearest neighbours. It is very plausible that forces originating in the repulsive potential should distort the electron distribution; the distortion should rather be regarded as the source of the repulsive potential. One way to visualize this and to make calculations is by using the shell model. In its simplest form the polarizability of the positive ion is neglected, so it is represented by a rigid ion carrying a charge Ze, while the negative ion is represented by a rigid core surrounded by a massless shell. The shell carries a charge $Ye (Y < 0)$ and is bound to the core with a force constant k. The total charge on core plus shell is $-Ze$. The short range interaction with nearest neighbour positive ions is through the shell, as indicated in Fig. 7.8. It is readily seen that while the polarizability of the free ion is $(Ye)^2/k\epsilon_0$, its polarizability in the crystal is

$$\alpha_2 = \frac{(Ye)^2}{(k + \beta)\epsilon_0} \tag{7.34}$$

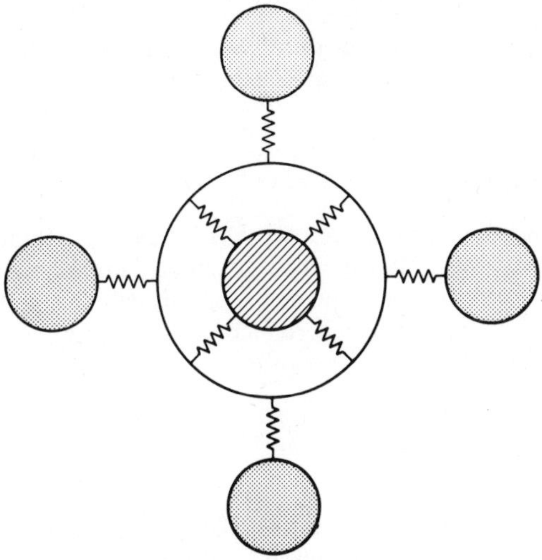

Figure 7.8 Schematic representation of the shell model of an ionic crystal. The polarizable negative ion is shown, and four adjacent positive ions.

The 'effective' charge of the negative ion when it is displaced is found to be given by

$$-Z' = -Z - \frac{Y\beta}{k+\beta}$$

that is

$$|Z'| < |Z| \tag{7.35}$$

since forces from its neighbours prevent it displacing as a rigid unit. On carrying through the calculation it is found that Z' replaces Z in Equ. 7.33. The additivity rule for the polarizabilities, described in §7.2, is not invalidated; since in practice it is found that $k \sim 10\beta$, the variation of β from one alkali halide to another produces only minor variations in α_2 (Equ. 7.34). In this way it is possible to account satisfactorily for the dielectric properties and the frequencies $\omega_j(\mathrm{q})$ (see Fig. 4.4 for KBr). It is the ionic charge $Z = 1$ which determines the cohesive energy and the elastic constants. Various elaborations of the model have been made (ref. 7.3) to take account of the polarizabilities of both ions, short range interaction between next nearest neighbours etc.

Although the shell model appears rather a crude mechanical model, it is simply a visual aid and a convenient way of formulating the equations. The same equations can be obtained without using concepts other than ionic charges and electronic dipoles. The latter are produced by, and in turn influence, both the field and the configuration of surrounding ions and electronic dipoles.

7.5 Infra-red spectroscopy

For most crystals the dielectric constant $\epsilon(\omega)$ does not exhibit any interesting variations in the frequency range in which it can be determined by measuring the capacity of a capacitor. The frequency ω_T is usually in the infra-red region of the electromagnetic spectrum. The reflectivity of a crystal at normal incidence is given by

$$R = \frac{|n-1|^2}{|n+1|^2} = \frac{(n'-1)^2 + (n'')^2}{(n'+1)^2 + (n'')^2} \tag{7.36}$$

where the 'optical constants' n' and n'' are related to the real and imaginary parts of ϵ by Equ. 7.10. Our discussion so far has given a dielectric constant which has only a real part. This results from reliance on the harmonic approximation. If we modify Equ. 7.23 for the rigid ion model so that it becomes

$$m\left(\frac{\partial^2 P}{\partial t^2} + 2\Gamma_T \frac{\partial P}{\partial t}\right) = \frac{(Ze)^2}{v}\left(E + \frac{P}{3\epsilon_0}\right) - \beta P \tag{7.37}$$

the solution for the dielectric constant, when E varies as $\exp(-i\omega t)$, is

$$\epsilon(\omega) - 1 = \frac{(Ze)^2/mv\epsilon_0}{\omega_T^2 - \omega^2 - 2i\Gamma_T\omega} = \frac{\omega_T^2(\epsilon(0) - 1)}{\omega_T^2 - \omega^2 - 2i\Gamma_T\omega} \tag{7.38}$$

When electronic polarizability is included the result is

$$\epsilon(\omega) - \epsilon(\infty) = \frac{\omega_T^2(\epsilon(0) - \epsilon(\infty))}{\omega_T^2 - \omega^2 - 2i\Gamma_T\omega} \tag{7.39}$$

The real and imaginary components of $\epsilon(\omega)$ are sketched in Fig. 7.9, using Equ. 7.39. Note that when $2\Gamma_T$ is relatively small, the full width at half-maximum of the peak in $\epsilon''(\omega)$ is just $2\Gamma_T$. This 'damping constant' (the factor 2 is traditional) introduces a mechanism by which energy can 'leak out' from the transverse optic mode into other modes.

Measurements on a number of diatomic crystals have given results for the reflectivity in fairly good agreement with Equ. 7.39 for the dielectric constant. If the damping constant were zero the reflectivity would be 100% between ω_T and ω_L. Fig. 7.10 shows the reflectivity calculated for $\omega_T/2\pi = 1{\cdot}098 . 10^{13}$ sec^{-1}, $\epsilon(0) = 10{\cdot}182$, $\epsilon(\infty) = 8{\cdot}457$ and $2\Gamma_T = 0{\cdot}003\omega_T$,

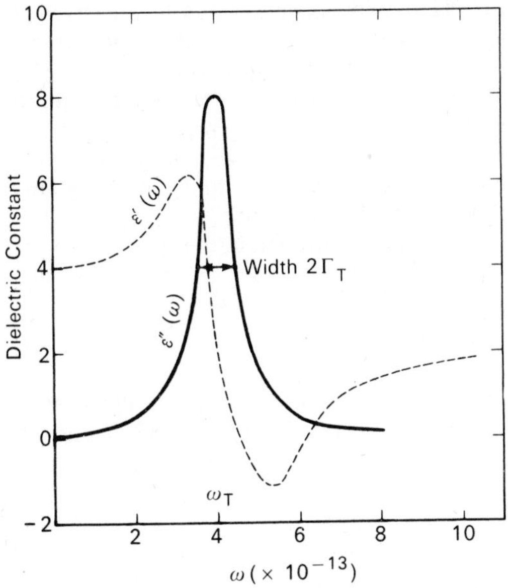

Figure 7.9 The values of $\epsilon'(\omega)$ and $\epsilon''(\omega)$, drawn for $\epsilon(0) = 4$, $\epsilon(\infty) = 2$, $\omega_T = 4.10^{13}$ s^{-1} and $2\Gamma_T = 10^{13}$ s^{-1}.

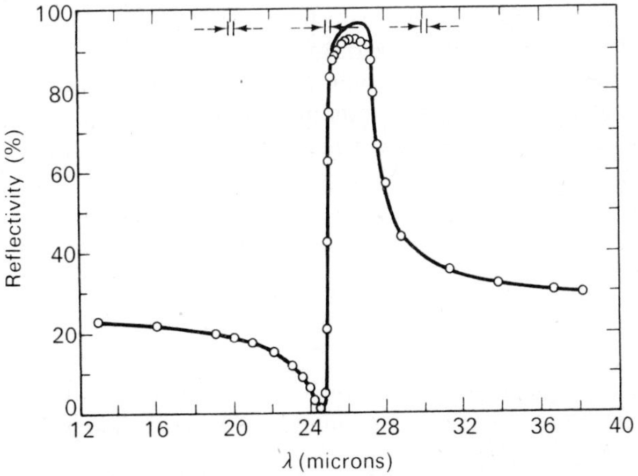

Figure 7.10 A comparison of calculated and measured values of the reflectivity of GaP. The abscissa is wavelength, $2\pi c/\omega$. The experimental resolution is indicated at three different wavelengths. (After Kleinman and Spitzer, 1960, *Phys. Rev.* **118**, 110.)

values appropriate to GaP. Experimental points are also shown (ref. 7.4); the agreement is good. Thin specimens show a maximum in infra-red absorption at the frequency ω_T. The relatively thick specimens of GaP used showed considerable absorption over a range of frequency. This results from processes in which a photon is absorbed and two phonons are created, or one is created and one destroyed. Such processes also affect the reflectivity but in practice by too small an amount to spoil the agreement with the simple theory given above, which of course does not take them into account.

When the above values are substituted in the equation

$$\epsilon(0) - \epsilon(\infty) = \frac{(\epsilon(\infty) + 2)^2 (Z'e)^2}{9v\epsilon_0 m\omega_T^2} \tag{7.40}$$

the value found for Z' is 0·58. This however tells us practically nothing about the ionic charge of GaP, for we note from Equ. 7.35 that it is possible in principle to have $Z' \neq 0$ when $Z = 0$.

7.6 Light scattering by crystals

So far we have considered a situation where, in terms of phonons, a photon is absorbed and a phonon is created with conservation of energy and pseudo-momentum (absorption of radiation), or where a photon collides elastically with the crystal without creating an excitation in it (reflection of radiation). It is also possible for photons to be scattered by phonons. The process is similar to the inelastic scattering of neutrons which we considered in §5.2. For historical reasons, scattering of light by acoustic modes is called Brillouin scattering, while scattering by optic modes is referred to as Raman scattering. The scattering cross section, as for neutron scattering, is greatest when one phonon is involved, and the conservation laws which apply are the appropriate modification of equations 5.16, namely

$$\mathbf{K} \equiv \mathbf{k}_1 - \mathbf{k}_0 = \mathbf{K}_h \pm \mathbf{q} \tag{7.41a}$$

and

$$\hbar\omega \equiv \hbar(\omega_0 - \omega_1) = \pm \hbar\omega_j(\mathbf{q}) \tag{7.41b}$$

where ω_0, $\mathbf{k}_0$ are respectively the frequency and wavevector of the incident light, and subscript 1 refers to scattered light. Frequency and wavevector are related by

$$\omega_0 = \frac{c}{\sqrt{\epsilon(\infty)}} k_0 \tag{7.42}$$

However, the wavelength of visible light is large compared with the unit cell dimension so that there is a solution of Equ. 7.41a only for $K_h = 0$ and for values of q very small compared with $2\pi/a$.

In Brillouin scattering, acoustic modes of long wavelength produce strain and thereby modulate the dielectric constant (optical refractive index) of the medium. The wavevector is such that

$$\omega_j(\mathbf{q}) = q u_j(\mathbf{q}) \tag{7.43}$$

where the velocity $u_j(\mathbf{q})$ is determined by the elastic constants (§5.5) for each of the three acoustic branches. To satisfy the equations with ω_0 and ω_1 much

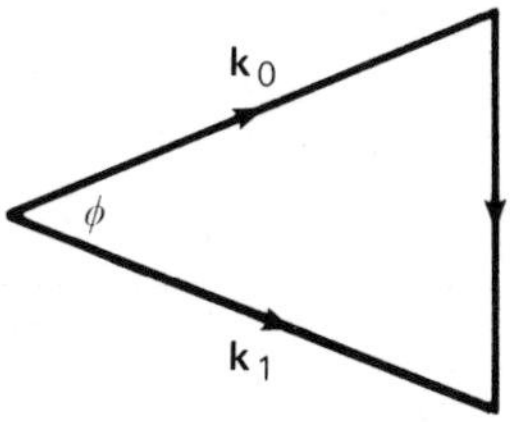

Figure 7.11 Relation between wavevectors $\mathbf{k}_0$ and $\mathbf{k}_1$ of incident and scattered light and the phonon wavevector $\mathbf{q}$, in Brillouin scattering.

greater than $\omega_j(\mathbf{q})$ requires $k_0 \triangleq k_1$, i.e. the scattering is almost elastic. The relation between $\mathbf{k}_0$, $\mathbf{k}_1$ and $\mathbf{q}$ is illustrated in Fig. 7.11, and leads to

$$q \triangleq 2k_0 \sin\frac{\phi}{2} \tag{7.44}$$

which is of the same form as Bragg's law for X-ray reflection, with $2\pi/k_0$ for λ and $2\pi/q$ for d. Combining equations 7.43 and 7.44, and using Equ. 7.42 gives

$$\omega_j(\mathbf{q}) = 2u_j(\mathbf{q}) \frac{\sqrt{\epsilon(\infty)}\,\omega_0}{c} \sin\frac{\phi}{2} \tag{7.45}$$

The maximum value of q is approximately $2k_0$ and the corresponding maximum value of $\omega_j(\mathbf{q})$ is in practice of the order $2\pi \,.\, 10^{10}\ \text{sec}^{-1}$. This is in a range between those frequencies accessible to investigation by ultrasonic and neutron techniques. Equ. 7.45 gives the frequency of the scattered light as

$$\omega_1 = \omega_0 \left(1 \pm \frac{2u_j(\mathbf{q})\sqrt{\epsilon(\infty)}}{c} \sin\frac{\phi}{2}\right) \tag{7.46}$$

The velocities of acoustic modes, and hence the elastic constants, can therefore be deduced from the change in frequency of the light scattered in particular directions. The intensity of scattering depends on the change of refractive index with the strain induced by the mode involved. It thus involves elements of the tensor which gives the ratio of change of refractive index to strain. These are called the Pockels or elasto-optic coefficients. The experimental technique involves a laser as source and generally a Fabry–Perot spectrometer to analyse the scattered light. A typical Brillouin spectrum is shown in Fig. 7.12.

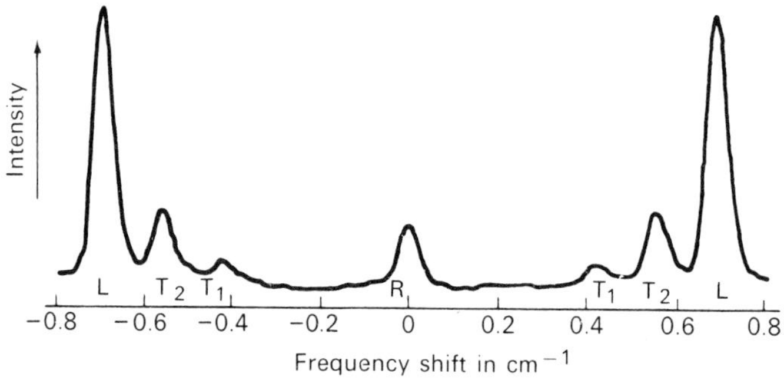

Figure 7.12 A Brillouin spectrum of quartz, showing scattering by all three acoustic modes. Frequency shifts $(\omega_1 - \omega_0)/2\pi$ are given in cm^{-1}, a unit used in spectroscopy which corresponds to a frequency of $3.10^{10}\ \mathrm{s}^{-1}$. (After Shapiro, Gammon and Cummins, 1966, *Appl. Phys. Lett.* 9, 157.)

In terms of classical physics, we can think of Raman scattering as coming about in the following way. If the polarizability of an atom depends on its displacement u, the dipole moment can be written as

$$p = \epsilon_0(\alpha + \alpha' u)E_l \tag{7.47}$$

Consequently if u varies with frequency $\omega_j(\mathbf{q})$ and E_l with the frequency ω_0 of the incident light, the dipole moment and the scattered light will have components of frequency

$$\omega_1 = \omega_0 \pm \omega_j(\mathbf{q}) \tag{7.48}$$

The second condition is the usual one for constructive interference of the scattered rays,

$$\mathbf{k}_1 - \mathbf{k}_0 = \mathbf{q} \tag{7.49}$$

The frequency $\omega_j(\mathbf{q})$ of an optic mode normally varies slowly with $\mathbf{q}$ for small values of q, and therefore the frequency change is practically independent of the angle of scattering, and corresponds closely to $q \to 0$. Not all optic modes of zero wavevector are 'Raman active'. The problem can be discussed succinctly using group theory, but the general rule apparent from the elementary discussion already given is that the polarizability of atoms must vary with the same frequency as that of the optic mode involved, for one-phonon scattering to be possible. When every atom of the structure is at a centre of symmetry there can be no term linear in the displacement in Equ. 7.47 and all modes are 'Raman inactive'. Thus alkali halides show only a continuous spectrum with no peaks at $\omega_1 - \omega_0 = \omega_T$ or ω_L, the continuous and relatively weak spectrum originating in processes involving two phonons, and which depend on anharmonic

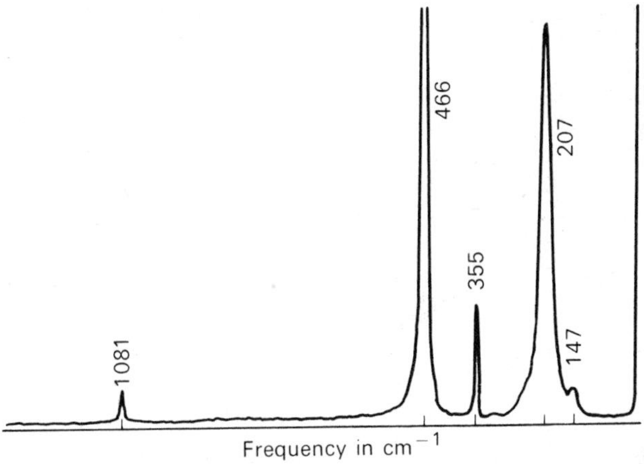

Figure 7.13 A Raman spectrum of quartz. Frequencies are given in cm^{-1}. The small peak at 147 cm^{-1} is believed to result from anharmonic effects. (After Shapiro, O'Shea and Cummins, 1967, *Phys. Rev. Lett.* **19**, 361.)

effects. The formula for the one-phonon scattering cross section resembles that for inelastic neutron scattering, Equ 5.13. In particular the cross section for scattering with loss of energy involves $(\bar{n}_j(\mathbf{q}) + 1)\hbar\omega_j(\mathbf{q})$ while that for scattering with gain of energy involves $\bar{n}_j(\mathbf{q})\ \hbar\omega_j(\mathbf{q})$. If the incident light is polarized in the x direction, and the scattered light in the y direction, the intensity of scattering depends on $(\partial\alpha_{xy}/\partial Q_j(\mathbf{q}))^2$ where α_{xy} is a component of the polarizability tensor and $Q_j(\mathbf{q})$ is the normal coordinate of the mode involved. Fig. 7.13 shows the Raman spectrum of quartz at 293 K for $\mathbf{k}_0$ and $\mathbf{k}_1$ at right angles to one another and to the c axis of the crystal. The polarization of both the incident and scattered light was parallel to the c axis. Only four of the numerous optic modes of zero wavevector scatter in this configuration.

Raman spectroscopy is generally capable of greater resolution in energy than is neutron spectroscopy, but the restriction to modes of zero wavevector in optically transparent materials makes it a less versatile technique.

References and suggestions for further reading

7.1 Tessman, J., Kahn, A., and Shockley, W. 1953, *Phys. Rev.* **92**, 890.
7.2 Daniels W. B. and Silverman, H. L., *Bulletin of the American Physical Society*, April 1968.
7.3 Cochran, W. 1971, *C.R.C. Critical Reviews in Solid State Sciences*, **2**, 1.
7.4 Kleinman, D. A. and Spitzer. W. 1960, *Phys. Rev.* **118**, 110.

The use of group theory in Raman spectroscopy has been reviewed by Loudon, R. 1964, *Adv. Phys.* **13**, 423.

8

Anharmonic Effects

8.1 Coefficient of Expansion

We have several times remarked in earlier chapters on phenomena that cannot be accounted for without going beyond the harmonic approximation. The increase of the specific heat beyond the value $3Nk_B$ with increasing temperature, the existence of thermal resistance and coefficient of expansion, and the occurrence of multi-phonon processes in Raman scattering are examples.

For most materials anharmonic effects are relatively small, and can be treated as a perturbation, becoming more important as the temperature increases. The change of volume that accompanies change of temperature affects physical properties, apart from any direct effect of the temperature change. When a crystal expands, its elastic energy increases, but this may be compensated by a decrease in free energy resulting from a lowering of the phonon frequencies. The equilibrium volume is that which minimizes the free energy. We quote without proof the result (ref. 8.1) that at a relatively high temperature the free energy of the lattice vibrations is

$$F_{lv} = k_B T \sum_{\mathbf{q}j} \ln\left(\frac{\hbar\omega_j(\mathbf{q})}{k_B T}\right) \tag{8.1}$$

When a cubic crystal is expanded from volume V to $V + \Delta V$, the increase in elastic energy is readily shown to be

$$U_{el} = \frac{(\Delta V)^2}{2\kappa V} \tag{8.2}$$

where κ is the compressibility (see Equ. 5.24). The relevant free energy is therefore

$$F = k_B T \sum_{\mathbf{q}j} \ln\left(\frac{\hbar\omega_j(\mathbf{q})}{k_B T}\right) + \frac{(\Delta V)^2}{2\kappa V} \tag{8.3}$$

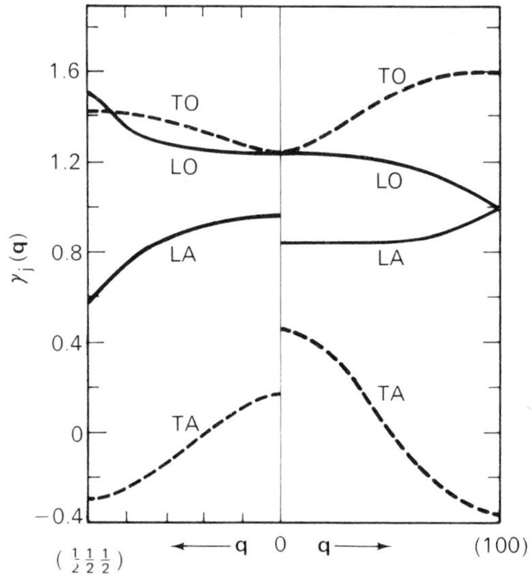

Figure 8.1 Calculated values of the Gruneisen parameter $\gamma_j(\mathbf{q})$ of germanium, for $\mathbf{q}$ parallel to [100] and [111] directions. (After Dolling and Cowley, 1966, *Proc. Phys. Soc.* 88, 463.)

Putting $(\partial F/\partial V)_T = 0$ we obtain

$$\frac{\Delta V}{V} = -\kappa k_B T \sum_{\mathbf{q}j} \frac{1}{\omega_j(\mathbf{q})}\left(\frac{\partial \omega_j(\mathbf{q})}{\partial V}\right) \tag{8.4}$$

Introducing

$$\gamma_j(\mathbf{q}) = -\frac{V}{\omega_j(\mathbf{q})}\left(\frac{\partial \omega_j(\mathbf{q})}{\partial V}\right) = -\frac{d \ln \omega_j(\mathbf{q})}{d \ln V} \tag{8.5}$$

we have for the volume strain

$$\frac{\Delta V}{V} = \frac{\kappa k_B T}{V} \sum_{\mathbf{q}j} \gamma_j(\mathbf{q}) \tag{8.6}$$

$\gamma_j(\mathbf{q})$ is called the Gruneisen parameter of the mode. If one begins from the expression for F_{lv} valid at a general temperature, which is (ref. 8.1)

$$F_{lv} = k_B T \sum_{\mathbf{q}j} \ln\left(2 \sinh\left(\frac{\hbar\omega_j(\mathbf{q})}{2k_B T}\right)\right) \tag{8.7}$$

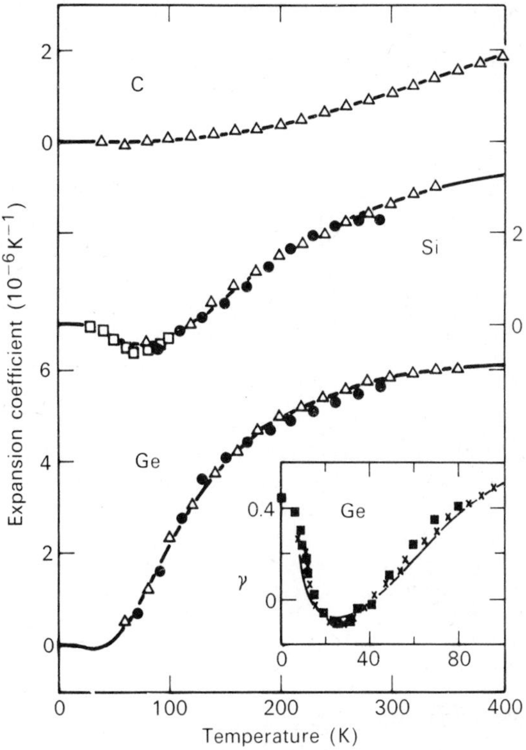

Figure 8.2 Comparison of measured and calculated values of the coefficients of expansion for diamond, silicon and germanium. The inset gives the mean Gruneisen parameter of germanium at low temperatures, showing negative values. (After Dolling and Cowley, 1966, *Proc. Phys. Soc.* 88, 463.)

one obtains at the expense of more algebra the result

$$\frac{\Delta V}{V} = \frac{\kappa}{V}\sum_{\mathbf{q}j}(\bar{n}_j(\mathbf{q}) + \tfrac{1}{2})\hbar\omega_j(\mathbf{q})\gamma_j(\mathbf{q}) \tag{8.8}$$

which reduces to Equ. 8.6 when the mean energy of any mode can be replaced by $k_B T$. Hence the coefficient of volume expansion is given by

$$\beta = \frac{d}{dT}\left(\frac{\Delta V}{V}\right) = \frac{\kappa}{V}\sum_{\mathbf{q}j} C_j(\mathbf{q})\gamma_j(\mathbf{q}) \tag{8.9}$$

where $C_j(\mathbf{q})$ is just the contribution of a mode to the specific heat, (§6.4). If the dependence of frequency on volume can be taken to be the same for all modes, so that $\gamma_j(\mathbf{q}) = \gamma$, a constant, Equ. 8.9 becomes

$$\beta = \kappa\gamma C \tag{8.10}$$

where C is the specific heat of unit volume. This is the Gruneisen relation. Although there is good reason to believe that $\gamma_j(\mathbf{q})$ will not usually be the same for all modes, for many materials the coefficient of expansion follows closely the variation of specific heat. For most materials the value of γ is between 1 and 2. Germanium and silicon however, have a negative coefficient of expansion at low temperatures, indicating that $\gamma_j(\mathbf{q})$ is negative for certain acoustic modes, which dominate β at low temperatures because of their relatively large value of $C_j(\mathbf{q})$.

In the harmonic approximation $\beta = 0$ since $\omega_j(\mathbf{q})$ is independent of volume. If the interatomic potential is known, as in an alkali halide, or can be approximated by a $\phi(r)$ involving a few parameters, the $\omega_j(\mathbf{q})$ can be calculated in the harmonic approximation, but for any interatomic separation, i.e. volume, using the theory outlined in chapter 4. In this way values of $\gamma_j(\mathbf{q})$ and of β have been calculated (ref. 8.2) for diamond, silicon and germanium, with results shown in Figs. 8.1 and 8.2 The agreement with the measured coefficients of expansion is good.

8.2 Normal modes and damping constants

In discussing approximate theories of thermal conductivity, and the dielectric constant, it was necessary to assume that energy could be transferred from one mode to others. In Chapter 6 we assumed that the phonon occupation number decreases exponentially to its equilibrium value with a relaxation time $\tau_j(\mathbf{q})$, (Equ. 6.33). In Chapter 7 a damping constant was introduced. The solution of Equ. 7.37, for $E = 0$, which represents a free vibration of the polarization P, is that the amplitude decreases as $\exp(-\Gamma_T t)$. The energy of any mode, in this approximation, therefore decreases as $\exp(-2\Gamma_j(\mathbf{q})t)$. Since the energy is proportional to the occupation number (Equ. 4.12) these two approaches are at least consistent with one another, and we can identify $2\Gamma_j(\mathbf{q})$ with $(\tau_j(\mathbf{q}))^{-1}$. We shall see later however that this is an oversimplification.

The result which we had for the dielectric constant of a diatomic crystal, Equ. 7.39, can be generalized to apply to a cubic crystal with several atoms per unit cell, following the method described in §7.5. The result is

$$\epsilon(\omega) - \epsilon(\infty) = \sum \frac{A_j}{\omega_j^2(0) - \omega^2 - 2i\Gamma_j(0)\omega} \tag{8.11}$$

The sum is over all transverse optic modes for which $q \to 0$, of which the number is generally just $n - 1$, for n atoms per unit cell. A_j is a constant involving the eigenvectors and effective charges, and $2\Gamma_j(0)$ is the damping constant for the mode, regarded simply as a parameter to be determined by experiment. Evidently each mode is taken to be a damped simple harmonic oscillator, resonating at the frequency $\omega_j(0)$. The function $\epsilon''(\omega)$ consists of $n - 1$ peaks each of width $2\Gamma_j(0)$ centred on the frequency $\omega_j(0)$; qualitatively this is the shape of the infra-red absorption spectrum. For any mode, not necessarily having $q = 0$, we can define a 'response function' which in this approximation is

$$R_j(\mathbf{q}\omega) = \{\omega_j^2(\mathbf{q}) - \omega^2 - 2i\Gamma_j(\mathbf{q})\omega\}^{-1} \tag{8.12}$$

In this notation, the dielectric constant is given by

$$\epsilon(\omega) - \epsilon(\infty) = \sum_j A_j R_j(0\omega) \tag{8.13}$$

In Chapter 5 we had an expression based on the harmonic approximation, for the frequency analysed neutron scattering cross section, Equ. 5.13. When the modes are each characterized by a damping constant $2\Gamma_j(\mathbf{q})$, the δ-functions of Equ. 5.13 are replaced by Lorentzian peaks each of width $2\Gamma_j(\mathbf{q})$ in frequency:

$$S_1(\mathbf{K}\omega) = Nb^2 \exp(-2w) \sum_j \frac{(\mathbf{K}\,.\,\mathbf{e}_j(\mathbf{q}))^2 \hbar\omega_j(\mathbf{q})}{m\omega_j^2(\mathbf{q})} \times$$
$$\left[(\bar{n}_j(\mathbf{q}) + 1)\frac{\Gamma_j(\mathbf{q})}{(\omega_j(\mathbf{q}) - \omega)^2 + \Gamma_j^2(\mathbf{q})} + \bar{n}_j(\mathbf{q})\frac{\Gamma_j(\mathbf{q})}{(\omega_j(\mathbf{q}) + \omega)^2 + \Gamma_j^2(\mathbf{q})}\right] \tag{8.14}$$

The part of the expression in square brackets can be written in terms of the imaginary part of $R_j(\mathbf{q}\omega)$. Using the result

$$\pi\delta(x) = \lim_{\epsilon \to 0} \frac{\epsilon}{x^2 + \epsilon^2} \tag{8.15}$$

we see that Equ. 8.14 reverts to Equ. 5.13 for $\Gamma_j(\mathbf{q}) \to 0$. An expression similar to Equ. 8.14 applies to the spectrum of light scattered by a crystal; the finite widths of the peaks are apparent in practice in Figs. 7.12 and 7.13. In a neutron spectrum the intrinsic widths are often masked by the instrumental resolution not being sufficiently great. To summarize therefore, if we assume that the amplitude of each mode decreases exponentially with time, the imaginary part of the response function of the mode has a peak at $\omega_j(\mathbf{q})$ of width $2\Gamma_j(\mathbf{q})$.

Experimentally, the peaks corresponding to transverse optic modes for which $q = 0$ are seen in $\epsilon''(\omega)$, and peaks corresponding to a wavevector $\mathbf{q}$ such that

$$\mathbf{K} \pm \mathbf{q} = \mathbf{K}_h \tag{8.16}$$

(see §5.2) are seen in $S_1(\mathbf{K}\omega)$, which determines the scattering cross section for radiation.

8.3 Frequency widths and shifts

The theory of the dynamics of a crystal in which anharmonic effects are weak (ref. 8.3) gives the expression for the response function as

$$R_j(\mathbf{q}\omega) = \{\omega_j^2(\mathbf{q}) - \omega^2 + 2\omega_j(\mathbf{q})(\Delta_j(\mathbf{q}\omega) - i\Gamma_j(\mathbf{q}\omega))\}^{-1} \tag{8.17}$$

Since $\Delta_j(\mathbf{q}\omega)$ and $\Gamma_j(\mathbf{q}\omega)$ are assumed small compared with the frequency in the harmonic approximation, $\omega_j(\mathbf{q})$, this may equally well be written

$$R_j(\mathbf{q}\omega) = \{(\omega_j(\mathbf{q}) + \Delta_j(\mathbf{q}\omega) - i\Gamma_j(\mathbf{q}\omega))^2 - \omega^2\}^{-1} \tag{8.18}$$

and we see that it corresponds to a time variation of the normal coordinate or amplitude for the mode $\mathbf{q}j$, of the form

$$Q_j(\mathbf{q}) \sim \exp(-i(\omega_j(\mathbf{q}) + \Delta_j(\mathbf{q}\omega))t)\exp(-\Gamma_j(\mathbf{q}\omega)t) \tag{8.19}$$

Thus the effective frequency of the mode is $\omega_j(\mathbf{q}) + \Delta_j(\mathbf{q}\omega)$ and is shifted by an amount which depends on the frequency ω with which the crystal is 'probed', and the damping constant $\Gamma_j(\mathbf{q}\omega)$ similarly depends on ω. As we shall see in more detail later, $\Delta_j(\mathbf{q}\omega)$ and $\Gamma_j(\mathbf{q}\omega)$ depend on temperature; even at $T = 0$ the existence of zero-point energy (see §6.1) prevents the effective frequency from being exactly equal to $\omega_j(\mathbf{q})$. The dielectric constant is still given by Equ. 8.13, but with $R_j(0\omega)$ now given by Equ. 8.17. The neutron scattering cross section is obtained from Equ. 8.14 by replacing $\omega_j(\mathbf{q})$ by $\omega_j(\mathbf{q}) + \Delta_j(\mathbf{q}\omega)$ and $\Gamma_j(\mathbf{q})$ by $\Gamma_j(\mathbf{q}\omega)$.

A resonance in the response function and a peak in its imaginary component will be determined by the condition

$$\omega_j^2(\mathbf{q}) - \omega^2 + 2\omega_j(\mathbf{q})\Delta_j(\mathbf{q}\omega) = 0 \tag{8.20}$$

The frequency satisfying this condition will be denoted $\widetilde{\omega}_j(\mathbf{q})$ and referred to as the quasi-harmonic frequency, and $\Gamma_j(\mathbf{q}\omega)$ evaluated at $\omega = \omega_j(\mathbf{q})$ will be similarly denoted $\widetilde{\Gamma}_j(\mathbf{q})$. The advantage of introducing these quantities is that it is often a reasonable approximation, at least in the neighbourhood of the resonances of the response function, to take it to be given

$$R_j(\mathbf{q}\omega) = \{\tilde{\omega}_j^2(\mathbf{q}) - \omega^2 - 2i\tilde{\Gamma}_j(\mathbf{q})\omega\}^{-1} \tag{8.21}$$

We have in effect validated the form of the response function, Equ. 8.12, which we wrote down by analogy with the 'classical' expression for $\epsilon(\omega)$,

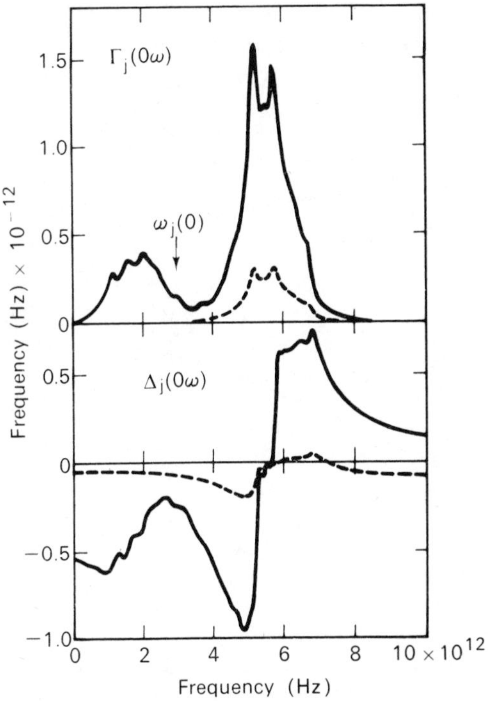

Figure 8.3 Calculated values of $\Delta_j(0\omega)$ (lower half) and of $\Gamma_j(0\omega)$ for KBr at temperatures of 5 K (dotted lines) and 300 K (full lines). (After Cochran and Cowley, 1967, *Handbuch der Physik* **25**/2a, 59, Springer Verlag, Berlin (Fig. 33, p. 119).)

Equ. 8.11, except that the harmonic frequency $\omega_j(\mathbf{q})$ has been replaced by the quasi-harmonic frequency $\tilde{\omega}_i(\mathbf{q})$. From the point of view of experiment, we see that $\tilde{\omega}_j(\mathbf{q})$ and $\tilde{\Gamma}_j(\mathbf{q})$ are to be determined from the measured positions and half-widths of the peaks in $\epsilon''(\tilde{\omega})$ or in $S_1(\mathbf{K}\omega)$.

The way in which $\Delta_j(\mathbf{q}\omega)$ and $\Gamma_j(\mathbf{q}\omega)$ depend on frequency, on temperature, and on the form of the interatomic potential, is outlined in the next section. An indication of their behaviour is given in Fig. 8.3 which shows the results of calculations of $\Delta_j(0\omega)$ and $\Gamma_j(0\omega)$ for the transverse optic mode (with $q = 0$) in KBr, at two temperatures. At high temperatures, $T > \theta_D$, both functions become proportional to temperature. At 5 K anharmonic effects are predicted to be small but not negligible, $\tilde{\omega}_j(0)$ is less than $\omega_j(0)$ by about 2%. Fig. 8.4

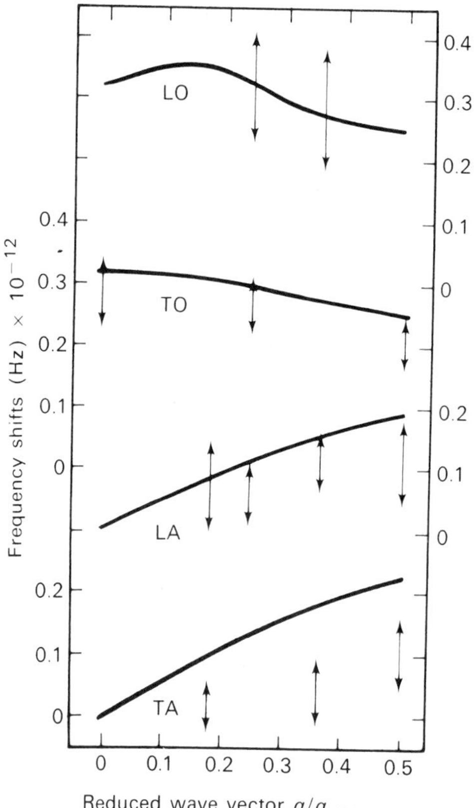

Figure 8.4 Comparison of claculated and measured increases of $\tilde{\omega}_j(\mathbf{q})$ when the temperature is lowered from 300 K to 90 K, for KBr. The wavevector **q** is in the [111] direction. LA indicates the longitudinal acoustic branch etc. (After Cowley, 1968, *Rep. Prog. Phys.* **31**, 123.)

shows a comparison between the calculated and measured changes of $\tilde{\omega}_j(\mathbf{q})$ between 400 K and 90 K, for KBr. Although the theory does not always give results in quantitative agreement with observation, this is largely due to our inadequate knowledge of the interatomic forces and not to deficiencies in the theory.

8.4 The influence of the interatomic potential

If we may take the potential energy of a crystal to be the sum of potentials between pairs of atoms, the harmonic frequencies $\omega_j(\mathbf{q})$ involve second derivatives of the interatomic potential, evaluated at the equilibrium interatomic distances, (at $T = 0$ and in the absence of zero-point energy, we may now add), as we saw in §4.1. Anharmonic effects are present to the extent that higher derivatives are appreciable. The detailed theory is beyond the scope of this book, and we do no more than quote some of the results which apply to a simple model, (ref. 8.4). The crystal is taken to be face-centred cubic with one atom in the primitive unit cell, with an interatomic potential $\phi(r)$ which extends only to nearest neighbours. There is then only one force constant, $\phi''(r_0)$, and the normal mode frequencies are given by

$$\omega_j(\mathbf{q}) = \left(\frac{2\phi''(r_0)}{m}\right)^{1/2} \lambda_j(\mathbf{q}) \tag{8.22}$$

where r_0 is the distance between nearest neighbours and $\lambda_j(\mathbf{q})$ is a dimensionless quantity which is independent of $\phi''(r_0)$ and of m. For example, when $\mathbf{q}$ is parallel to the [100] direction

$$\left.\begin{aligned} \lambda_1(\mathbf{q}) = \lambda_2(\mathbf{q}) &= \sqrt{2}\sin(\tfrac{1}{4}qa) \\ \lambda_3(\mathbf{q}) &= 2\sin(\tfrac{1}{4}qa) \end{aligned}\right\} \tag{8.23}$$

for transverse and longitudinal modes respectively. From Equ. 8.22 and the definition of the Gruneisen parameter (Equ. 8.5) it follows that the latter is given by

$$\gamma_j(\mathbf{q}) = -\frac{1}{3}\frac{d\ln\omega_j(\mathbf{q})}{d\ln r_0} = \frac{-r_0\phi'''(r_0)}{6\phi''(r_0)} \tag{8.24}$$

It is the same for all modes, so using the result that the energy of the crystal is

$$\bar{U} = \sum_{\mathbf{q}j} (\bar{n}_j(\mathbf{q}) + \tfrac{1}{2})\hbar\omega_j(\mathbf{q}) \tag{8.25}$$

(see §6.1) we have from Equ. 8.8 that the thermal strain is given by

$$\frac{\Delta V}{V} = \frac{\kappa\gamma\bar{U}}{V} \tag{8.26}$$

The shift in frequency of a mode can be written as the sum of three contributions

$$\Delta_j(q\omega) = \Delta_j^{(0)}(q) + \Delta_j^{(1)}(q) + \Delta_j^{(2)}(q\omega) \tag{8.27}$$

It will be noticed that two of these are in fact independent of ω. The first $\Delta_j^{(0)}(q)$, takes into account the expansion of the crystal and would be absent if the crystal were constrained to constant volume. It is readily evaluated as

$$\Delta_j^{(0)}(q) = \frac{\partial \omega_j(q)}{\partial V}\, \Delta V = -\omega_j(q)\kappa\gamma^2\, \frac{\bar{U}}{V} \tag{8.28}$$

In the high temperature limit it is given by

$$\Delta_j^{(0)}(q) = \frac{-k_B T(\phi'''(r_0))^2\, \omega_j(q)}{8(\phi''(r_0))^3} \tag{8.29}$$

We see that $\Delta_j^{(0)}(q)$ is always negative.

The second term which is independent of ω depends on quartic anharmonicity. The general expression for it is

$$\Delta_j^{(1)}(q) = \frac{\hbar}{4N\omega_j(q)} \sum_{q_1 j_1} \frac{(\bar{n}_1 + \frac{1}{2})}{\omega_1}\, \Phi_4(-qj;\, qj;\, q_1 j_1;\, -q_1 j_1) \tag{8.30}$$

Here ω_1 is an abbreviation for $\omega_{j_1}(q_1)$ and $\bar{n}_1$ for the corresponding occupation number. The quantity Φ_4 involves fourth derivatives of the interatomic potential; it also depends on the frequencies and eigenvectors of the modes qj and $q_1 j_1$. It is a special case of a more general function in which the sum of the four wavevectors involved is always zero. For our simple model the sum in Equ. 8.30 can be evaluated analytically in the high temperature limit, to give

$$\Delta_j^{(1)}(q) = \frac{k_B T\phi''''(r_0)\omega_j(q)}{8(\phi''(r_0))^2} \tag{8.31}$$

The general expressions for $\Delta_j^{(2)}(q\omega)$ and for $\Gamma_j(q\omega)$ are somewhat similar, and we quote only the latter:

$$\Gamma_j(q\omega) = \frac{\pi\hbar}{16N\omega_j(q)} \sum \frac{|\Phi_3(-qj;\, q_1 j_1;\, q_2 j_2|^2}{\omega_1 \omega_2} \times$$
$$\{-(\bar{n}_1 + \bar{n}_2 + 1)\,\delta(\omega + \omega_1 + \omega_2) + (\bar{n}_1 + \bar{n}_2 + 1)\delta(\omega - \omega_1 - \omega_2)$$
$$-(\bar{n}_1 - \bar{n}_2)\delta(\omega - \omega_1 + \omega_2) + (\bar{n}_1 - \bar{n}_2)\,\delta(\omega + \omega_1 - \omega_2)\} \tag{8.32}$$

The sum in this expression is to be taken over all modes $\mathbf{q}_1 j_1$ and $\mathbf{q}_2 j_2$ such that

$$-\mathbf{q} + \mathbf{q}_1 + \mathbf{q}_2 = \mathbf{K}_h \tag{8.33}$$

where as before $\mathbf{K}_h$ is any vector of the reciprocal lattice, including zero. This condition indicates that in terms of phonons, $\Gamma_j(\mathbf{q}\omega)$ (and also $\Delta_j^{(2)}(\mathbf{q}\omega)$) depend on the break-up of the mode $\mathbf{q}j$ into two modes, and their recombination, with conservation of wavevector. This is shown schematically in Fig. 8.5; the topological properties of such 'phonon diagrams' are in fact used in evaluating the contribution which each anharmonic process makes to $\Gamma_j(\mathbf{q}\omega)$ and to $\Delta_j(\mathbf{q}\omega)$. Φ_3 involves third derivatives of the interatomic potential, and also depends on the frequencies and eigenvectors of the three modes $\mathbf{q}j$, $\mathbf{q}_1 j_1$, and $\mathbf{q}_2 j_2$. Even for our simple model the sum in Equ. 8.32 can only be evaluated by numerical methods; it is found that $\Gamma_j(\mathbf{q}\omega)$ and $\Delta_j^{(2)}(\mathbf{q}\omega)$ are proportional to $\dfrac{(\phi'''(r_0))^2}{(\phi''(r_0))^3}$, and to T at high temperature. The three contributions to $\Delta_j(\mathbf{q}\omega)$ (Equ. 8.27) are generally of the same order of magnitude. $\Delta^{(2)}$ is positive and

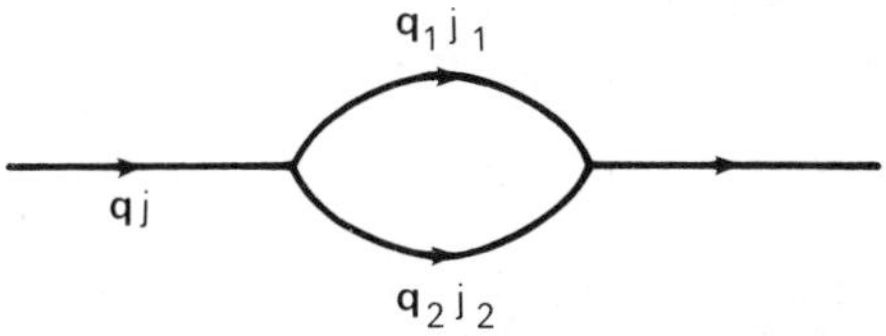

Figure 8.5 Phonon diagram for a third order anharmonic process.

may be cancelled by $\Delta^{(0)}$ and $\Delta^{(1)}$ so that $\tilde{\omega}_j(\mathbf{q})$ generally decreases as the temperature increases.

The condition (Equ. 8.33) was already quoted (Equ. 6.44) in discussing thermal conductivity, where the relaxation time in fact also involves $|\Phi_3|^2$. There however, we are concerned with scattering processes in which energy is conserved. The corresponding value of $\Gamma_j(\mathbf{q}\omega)$ is $\tilde{\Gamma}_j(\mathbf{q})$, for which $\omega = \tilde{\omega}_j(\mathbf{q}) \triangleq \omega_j(\mathbf{q})$ The δ-functions inside the brackets of Equ. 8.32 then ensure that all processes involving three phonons which contribute to $\tilde{\Gamma}_j(\mathbf{q})$ occur with conservation of energy. We have seen in §6.4 however that we must distinguish contributions to $2\tilde{\Gamma}_j(\mathbf{q})$ for which $K_h = 0$ (normal processes) from those for which $K_h \neq 0$ (umklapp processes), since these influence the thermal conductivity in very different ways. We cannot therefore simply equate $2\tilde{\Gamma}_j(\mathbf{q})$ with $(\tau_j(\mathbf{q}))^{-1}$, where $\tau_j(\mathbf{q})$ is the relaxation time introduced in §6.4.

References and suggestions for further reading

8.1 Wallace, D. C. 1972, *Thermodynamics of Crystals*, John Wiley and Sons, New York and London.
8.2 Dolling, G. and Cowley, R. A. 1966, *Proc. phys. Soc.* **88**, 463.
8.3 Cowley, R. A., *Rep. Prog. Phys.* 1968, **31**, 123.
8.4 Maradudin, A. A. and Fein, A. E. 1962, *Phys. Rev.* **128**, 2589.

9

Electrons and Phonons

9.1 Formal analogies

The dispersion relation $\omega_j(\mathbf{q})$ for phonons gives immediately the energy as a function of wavevector and clearly corresponds to the $E_j(\mathbf{k})$ relation for electrons, where $\mathbf{k}$ is the wavevector of the electron and j identifies the different energy levels in the reduced zone scheme. It is interesting that the $E_j(\mathbf{k})$ relation can be derived in a way which closely parallels the derivation of the $\omega_j(\mathbf{q})$ relation. We begin by writing the wave function of an electron in the crystal as a linear combination of single atom wave functions:

$$\psi(\mathbf{r}t) = \sum_l C_l(t)\psi_l(\mathbf{r}) \tag{9.1}$$

$\psi_l(\mathbf{r})$ is centred on the site l, C_l is the probability amplitude for the electron to be on the atom at site l. The equation which determines the wave function is then

$$i\hbar \frac{\partial C_l}{\partial t} = \sum_{l'} E_{ll'} C_{l'} \tag{9.2}$$

where the $E_{ll'}$ are real numbers. Equ. 9.2 can be derived from the Schrödinger equation, but it may be regarded as equally fundamental (see Feynman, (ref. 9.1)). Now consider a one-dimensional crystal, and assume that the probability amplitude C_l is 'coupled' only to nearest neighbours—this is called the tight binding approximation. Equ. 9.2 is then

$$i\hbar \frac{\partial C_l}{\partial t} = E_0 C_l - E_1 C_{l-1} - E_1 C_{l+1} \tag{9.3}$$

This is similar to Equ. 3.1 for atomic displacements in a linear crystal, except that we have $\partial/\partial t$ instead of $\partial^2/\partial t^2$ on the left side. By analogy with earlier results, the solution of Equ. 9.3 is

$$C_l = N^{-1/2} \exp i(kla - \omega(k)t) \tag{9.4}$$

On substituting this solution back into Equ. 9.3 we obtain

$$E(k) = \hbar\omega(k) = E_0 - 2E_1 \cos ka \tag{9.5}$$

This relation for the energy band is shown, for the first Brillouin zone, in Fig. 9.1. Obviously it repeats with the periodicity of the reciprocal lattice. The wave function $\psi(\mathbf{r}t)$ is now seen to have the form required by Bloch's theorem, on substituting the solution (Equ. 9.4) in Equ. 9.1. Periodic boundary conditions can be used to give the allowed values of k; as for q they are separated by an interval $2\pi/Na$. It is at first puzzling that we have obtained

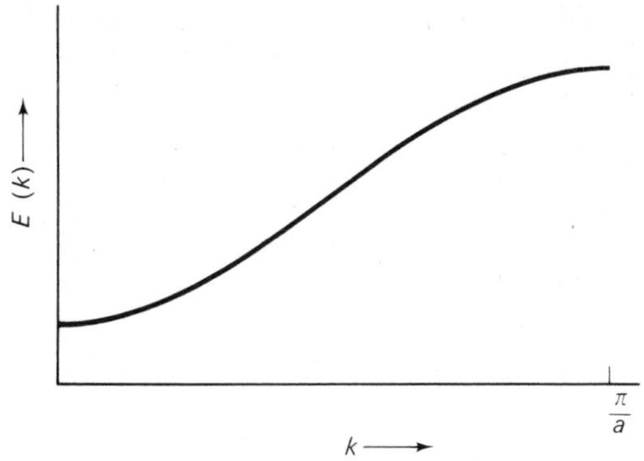

Figure 9.1 Electron energy as a function of wavevector, in one-dimension.

only one energy band. This is because Equ. 9.1 has not allowed for excited states of the atoms. It should be written more generally as

$$\psi(\mathbf{r}t) = \sum_{l\kappa} C_{l\kappa}(t)\psi_{l\kappa}(\mathbf{r})$$

where $\kappa = 0$ denotes the ground state, $\kappa = 1$ the first excited state etc. of the atom. Equation 9.2 is then replaced by an infinite set of equations, unlike the situation in lattice dynamics where the number of corresponding equations depends on the number of atoms per unit cell. There is thus no limit to the number of branches of the $E_j(\mathbf{k})$ relation, although in practice the equations may be solved by ignoring all but a few excited states.

Phonons obey Bose-Einstein statistics and there is no limit to the occupation number

$$n_j(\mathbf{q}) = \left\{\exp\left(\frac{\hbar\omega_j(\mathbf{q})}{k_B T}\right) - 1\right\}^{-1} \tag{9.6}$$

Electrons are governed by Fermi-Dirac statistics and

$$f_j(\mathbf{k}) = \left\{\exp\left(\frac{E_j(\mathbf{k}) - \mu}{k_B T}\right) + 1\right\}^{-1} \tag{9.7}$$

cannot exceed unity, although there are two states characterized by opposite spins for a particular $\mathbf{k}j$. Because of the unlimited number of branches the density of states function $G(E)$, unlike the phonon frequency distribution $G(\omega)$, has no upper limit. The density of occupied states $f(E)G(E)$ may however effectively end at a gap in $G(E)$ to give an insulator, or fall to zero in the middle of a band of allowed states to give a metal. These should be familiar results but it may be helpful to see them set out alongside one another.

9.2 Photons, phonons and electrons in germanium

When radiation of sufficient energy falls on an insulator or semiconductor, it can be absorbed with excitation of an electron from one band to another. The conditions which must apply are by now familiar

$$\left.\begin{aligned} E_f(\mathbf{k}_f) - E_i(\mathbf{k}_i) &= \hbar\omega_0 \\ \mathbf{k}_f - \mathbf{k}_i &= \mathbf{k}_0 + \mathbf{K}_h \end{aligned}\right\} \tag{9.8}$$

$\mathbf{k}_0$ and ω_0 refer to the radiation, subscripts i and f denote initial and final states of the electron. Radiation of appropriate energy has a wavevector much smaller than $2\pi/a$ (where as before a is the unit cell dimension) and therefore the second condition reduces to $\mathbf{k}_f \triangleq \mathbf{k}_i$. This is referred to as a direct transition. For a semiconductor with the maximum of the valence band and the minimum of the conduction band both at the centre of the Brillouin zone, at low temperatures the absorption coefficient is therefore a function of $\hbar\omega_0 - E_g$, being zero for $\hbar\omega_0 \leqslant E_g$ where

$$E_g = E_f(0) - E_i(0) \tag{9.9}$$

is the energy gap.

In certain crystals the maximum of the valence band and the minimum of the conduction band do not occur at the same value of k in the Brillouin zone. Fig. 9.2 shows part of the $E_j(\mathbf{k})$ relation for germanium. The maximum of the valence band is at the origin but the minimum of the conduction band is at $(2\pi/a)(\frac{1}{2}\frac{1}{2}\frac{1}{2})$. Equations 9.8 predict that there will now be no absorption of radiation for $\hbar\omega_0$ just greater than E_g, and at low temperatures this is indeed so. However, electrons can be excited by a process involving a phonon. With k_0

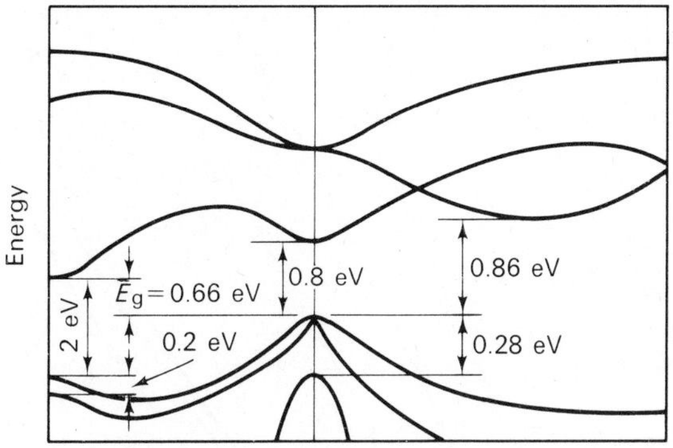

Figure 9.2 The energy–wavevector relation for electrons in germanium, for wavevector **k** in [100] and [111] directions. (From the article by Lax in *Semiconductors*, Ed. R. A. Smith, Academic Press, 1963 (report of conference organized by Italian Phys. Soc.), Fig. 13a, p. 255.)

and K_h taken to be zero for the same reason as before, the conservation equations are

$$E_f(\mathbf{k}_f) - E_i(\mathbf{k}_i) = \hbar\omega_0 \pm \hbar\omega_j(\mathbf{q}) \tag{9.10a}$$

$$\mathbf{k}_f - \mathbf{k}_i = \mathbf{q} \tag{9.10b}$$

At a low temperature, phonons with the requisite wavevector are not thermally excited, and the process is only possible with a minus sign in Equ. 9.10a, so that absorption begins at

$$\hbar\omega_0 = E_g + \hbar\omega_j(\mathbf{q}) \tag{9.11}$$

with creation of a phonon in the crystal, as well as excitation of an electron. At higher temperatures absorption begins at a lower photon energy, given by

$$\hbar\omega_0 = E_g - \hbar\omega_j(\mathbf{q}) \tag{9.12}$$

In this process a phonon is absorbed in the course of electron excitation.

Fig. 9.3 shows the square root of the absorption coefficient of *Ge* as a function of photon energy at various temperatures, (ref. 9.2). The major reason for the movement of the absorption to lower energy with increasing temperature is the gradual decrease of E_g, from 0·741 eV at 4 K to 0·664 eV at 291 K. At 4 K, onset of absorption at energy E_2 is followed by a sharp increase at a higher energy E_3, as reconstructed in more detail in Fig. 9.4a.

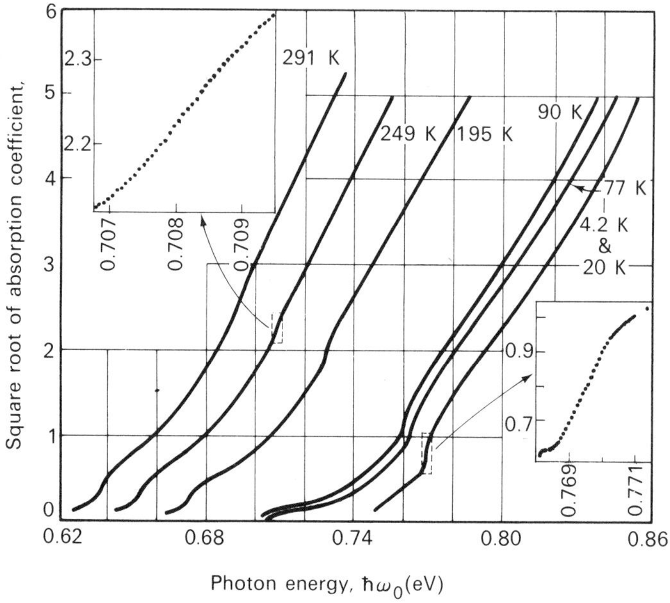

Figure 9.3 The square root of the absorption coefficient of radiation in germanium, at various temperatures. (After Macfarlane, McLean, Quarrington and Roberts, 1957, *Phys. Rev.* **108**, 1377.)

These correspond to processes in which a transverse acoustic phonon and a longitudinal acoustic phonon respectively with $\mathbf{q} = (2\pi/a)(\frac{1}{2}\frac{1}{2}\frac{1}{2})$, are produced. At higher temperatures there are four contributions to the absorption, as reconstructed in Fig. 9.4b. Two are the same as we have just identified; those beginning at E_0 and E_1 (Fig. 9.4b) are the corresponding processes in which a longitudinal and a transverse phonon respectively are absorbed, to assist the photon. (On the scale of Fig. 9.3 only the kinks in the curve at E_0 and E_3 are clearly visible at 195 K). The energy gap is given by $\frac{1}{2}(E_0 + E_3) = \frac{1}{2}(E_1 + E_2)$ at each temperature. The phonon energies are $\frac{1}{2}(E_2 - E_1)$ and $\frac{1}{2}(E_3 - E_0)$ for transverse and longitudinal modes respectively. The values deduced by MacFarlane *et al.* (ref. 9.2) were $0{\cdot}79.10^{-2}$ eV and $2{\cdot}80.10^{-2}$ eV, in good agreement with frequencies subsequently determined by neutron spectroscopy.

9.3 Scattering of electrons by phonons

The interactions of electrons and phonons in a metal is a topic which has occupied the attention of physicists for many years, not only because of its importance for determining the electrical and thermal conductivities of metals

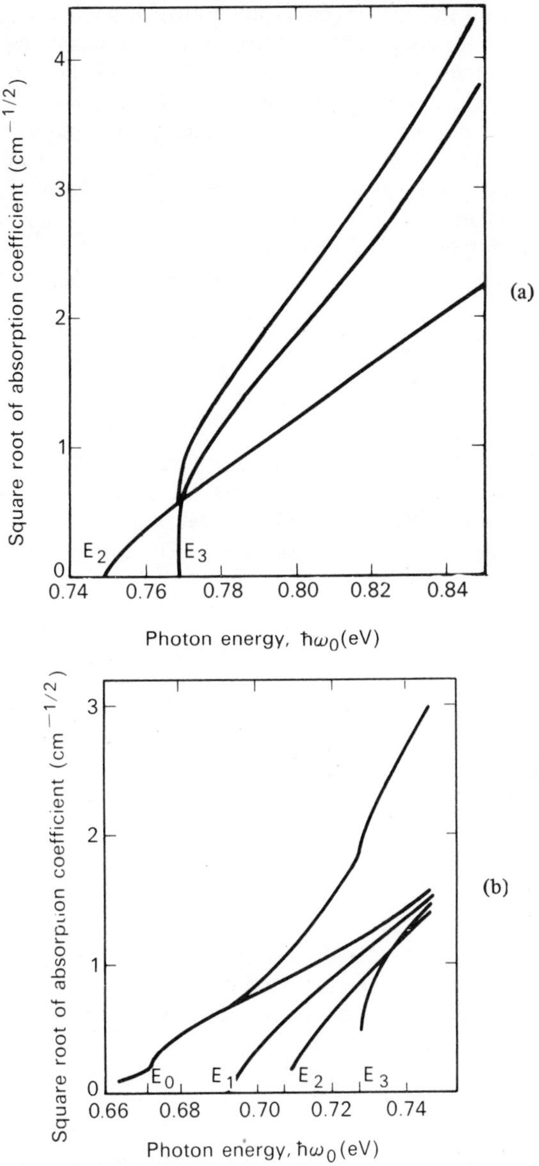

Figure 9.4 (a) The decomposition of the absorption into two components beginning at E_2 and E_3, at 4·2 K. (b) At 195 K there are four components, beginning respectively at E_0, E_1, E_2 and E_3. (After Macfarlane, McLean, Quarrington and Roberts, 1957, *Phys. Rev.* **108**, 1377.)

but also because of the challenging problems which the development of a theory presents.† The major complication is that as the ions move, the electron gas flows so as to screen the Coulomb interaction between the cores and individual electrons. The topic is considered in detail in other textbooks (ref. 9.3) and in accordance with our policy of trying to understand the physics of a process by considering the simplest situation, we shall discuss instead the intra-band scattering of electrons by acoustic lattice vibrations in a semiconductor such as silicon or germanium. There are then too few conduction electrons for screening to be important at such frequencies.

We take the conduction band to be parabolic, and define the effective electron mass by

$$E(k) = \frac{\hbar^2 k^2}{2m_e^*} \tag{9.13}$$

The electrical conductivity is determined primarily by the number of conduction electrons present, which varies as $T^{3/2} \exp(-E_g/2k_B T)$, but also by their mobility. We neglect the contribution of holes in the valence band, which could be taken into account in the same way. The conductivity σ_e is given by

$$\sigma_e = ne\mu_e \tag{9.14}$$

where n is the number of conduction electrons per unit volume and

$$\mu_e = \frac{e\tau}{m_e^*} \tag{9.15}$$

is their mobility. The relaxation time τ is given by

$$\tau = \frac{\lambda}{u_e} \tag{9.16}$$

where λ is the mean free path of an electron and u_e their average velocity. In an intrinsic semiconductor the conduction electrons are in the 'tail' of the Fermi–Dirac distribution and in effect obey classical statistics, so that an approximate value of u_e is given by

$$\tfrac{1}{2}m_e^* u_e^2 = \tfrac{3}{2}k_B T \tag{9.17}$$

For a conduction electron in Si, $m_e^* \sim 0{\cdot}2\, m_e$ and $u_e \sim 2{\cdot}6.10^7$ cm sec^{-1} at 293 K. The corresponding electron wavelength, $h/(m_e^* u_e)$, is about $1{\cdot}4.10^{-6}$ cm,

† It is also responsible for the remarkable property of superconductivity.

much too great to be Bragg reflected in the crystal. In a static crystal free of imperfections there would be no scattering mechanism and zero resistivity. While scattering by impurities and lattice defects is important at low temperatures, at room temperature scattering by phonons is the dominant process, just as for the thermal conductivity.

To calculate the mobility we proceed as follows. Consider a 'beam' of conduction electrons passing through the crystal. Thermal scattering attenuates the beam and by definition of mean free path its intensity diminishes by a factor $(1 - (dx/\lambda))$ in a distance dx. The number of unit cells contained in a slab of unit area and thickness dx is (dx/v). The fraction of the intensity scattered out is therefore $\sigma' dx/v$ where σ' is the scattering cross section per unit cell. It follows that

$$\frac{1}{\lambda} = \frac{\sigma'}{v} \tag{9.18}$$

The scattering conditions are the familiar equations

$$E(\mathbf{k}_1) - E(\mathbf{k}_0) = \hbar\omega_j(\mathbf{q}) \tag{9.19a}$$

and

$$\mathbf{K} = \mathbf{k}_1 - \mathbf{k}_0 = \mathbf{q} \tag{9.19b}$$

With $\hbar q \sim \hbar k_0 \sim (2m_e^* k_B T)^{1/2}$, and using

$$\omega_j(\mathbf{q}) = uq \tag{9.20}$$

where u is now the average velocity of a (longitudinal) acoustic wave, one finds $\hbar\omega_j(q) \sim (1/10)k_B T$. Thus the change of electron energy on scattering will usually be small compared with its initial energy, and we are justified in treating the scattering as elastic. In Chapter 5 we traced the main steps of a derivation which gave the intensity of quasi-elastic scattering of neutrons or of X-rays by lattice vibrations. The result was (Equ. 5.10)

$$S_1(\mathbf{K}) = Nf^2(K)\exp(-2w)\sum_j \frac{(\mathbf{K}\,.\,\mathbf{e}_j(\mathbf{q}))^2(\bar{n}_j(\mathbf{q}) + \frac{1}{2})\hbar\omega_j(\mathbf{q})}{m\omega_j^2(\mathbf{q})} \tag{9.21}$$

This gives the intensity (for unit incident intensity) scattered into unit solid angle in the direction of $\mathbf{k}_1$. $f(K)$ is now the appropriate scattering factor for *electrons*, which we shall discuss shortly. (Strictly speaking, since Si is a diatomic crystal, $f(K)$ should be replaced by a structure factor. This however turns out to be $2f(K)$ for an acoustic mode and zero for an optic mode, so we obtain the correct result by continuing to use the formula for a monatomic crystal, with v the volume containing one Si atom). In the situation we are considering, K is always sufficiently small that the Debye-Waller factor $\exp(-w)$ can be set equal to one and $f(K)$ can be replaced by $f(0)$. Using Equ. 9.19b we see that

$$(\mathbf{K}\,.\,\mathbf{e}_\mathrm{j}(\mathbf{q}))^2 = q^2 \tag{9.22}$$

for a longitudinal mode ($\mathbf{e}_\mathrm{j}(\mathbf{q})$ is a unit vector, §4.2) and is zero for a transverse mode. The electrons are scattered only by longitudinal modes. $(\bar{n}_\mathrm{j}(\mathbf{q}) + \frac{1}{2})\hbar\omega_\mathrm{j}(\mathbf{q})$ can be replaced by $k_\mathrm{B}T$. Equ. 9.21 then reduces to

$$S_1(K) = Nf^2(0)\left(\frac{k_\mathrm{B}T}{mu^2}\right) \tag{9.23}$$

where we have used Equ. 9.20. It follows that

$$N\sigma' = \int S_1(K)d\Omega = \frac{4\pi Nf^2(0)k_\mathrm{B}T}{mu^2} \tag{9.24}$$

where $d\Omega$ is an element of solid angle. We may approximate u^2 by the value appropriate to an acoustic wave travelling in the [100] direction, which is (§5.5)

$$u^2 = \frac{c_{11}}{D} = \frac{c_{11}v}{m} \tag{9.25}$$

Hence combining equations 9.18 and 9.24,

$$\frac{1}{\lambda} = \frac{4\pi k_\mathrm{B}Tf^2(0)}{c_{11}v} \tag{9.26}$$

All that remains is to decide on a value for $f(0)$. It can be shown (ref. 9.4) that for fast electrons scattered by an atom of atomic number Z,

$$f_\mathrm{a}(0) = \frac{\pi m_\mathrm{e}e^2 Z\langle r^2\rangle}{3h^2\epsilon_0} \tag{9.27}$$

where $\langle r^2 \rangle$ is the mean square radius of the atomic electron density. The values given in tables (ref. 9.4) for $f_a(0)$ are 6·0 Å for Si and 7·8 Å for Ge. To apply to electrons in the conduction band, we take

$$f(0) = f_a(0) \frac{m_e^*}{m_e} \tag{9.28}$$

and thus finally obtain for the mobility

$$\mu_e = \frac{ec_{11} v^2 m_e^2}{4\pi\sqrt{3}(k_B T)^{3/2} f_a^2(0)(m_e^*)^{5/2}} \tag{9.29}$$

The conduction energy band of Si is not in fact parabolic; the reciprocal effective mass tensor has two principal values of $(0{\cdot}19m_e)^{-1}$ and one of $(0{\cdot}98m_e)^{-1}$. This gives an average value of $(0{\cdot}26m_e)^{-1}$ for $(m_e^*)^{-1}$. Inserting this value in Equ. 9.29, with other values appropriate to Si ($c_{11} = 1{\cdot}66 \,.\, 10^{11}$ N m^{-2}, $v = 19{\cdot}9$ Å^3, $T = 293$ K) one finds $\mu_e = 1600$ cm^2 volt^{-1} sec^{-1}, in exact agreement with the measured value! Considering the approximations which have been made, the exact agreement is of course fortuitous. For Ge the calculated value is approximately twice the measured value of 3800 cm^2 volt^{-1} sec^{-1}.

We may conclude therefore that we do not go seriously wrong if we say that in a semiconductor the conduction electrons are scattered by phonons in just the same way as are X-rays or neutrons passing through the crystal, and it is this which determines the mobility at ordinary temperatures. In a pure metal at such temperatures the conductivity is again dominated by phonon scattering of the electrons but the latter are sufficiently numerous to make an important contribution to the scattering factor $f(0)$, which is thus not simply related to that of an isolated atom or ion.

References and suggestions for further reading

9.1 Feynman, R. P., Leighton, R. B., and Sands, M. 1965, *The Feynman Lectures on Physics*, Vol. 3, Addison–Wesley, New York.
9.2 MacFarlane, G. G., MacLean, T. P., Quarrington, J. E., and Roberts, V. 1957, *Phys. Rev.* **108**, 1377.
9.3 Ziman, J. M. 1960, *Electrons and Phonons*, Clarendon Press, Oxford, and also Smith, R. A. 1969, *Wave Mechanics of Crystalline Solids*, Chapman and Hall, London.
9.4 *International Tables for Crystal Structure Determination.* Vol. III, 1962, Ed. K. Lonsdale, Kynoch Press, Birmingham.

The Electronic Structures of Solids, by B. R. Coles and A. D. Caplin, is a companion volume in the series to which this book belongs, as is *The Electrical Properties of Solids*, by J. S. Dugdale. They deal in greater detail, and from a different point of view, with some of the topics considered in Sections 9.1 and 9.3 respectively.

For a thorough discussion of the theory of mobility in nonpolar semiconductors, see Lawaetz, P., 1969, *Phys. Rev.* **183**, 730.

10

Phase Transitions and Lattice Dynamics

10.1 General features of phase transitions

The most familiar examples of phase transitions are those from solid to liquid, or from liquid to gas. These, particularly the former, are transitions between phases which are dissimilar in atomic structure and as such are first order transitions. The Gibbs free energy (of a definite mass) is defined by

$$G = U - TS + pV \tag{10.1}$$

and is the same for two phases in contact at a particular pressure and temperature. The term first order refers to the fact that a *first* derivative of the free energy, such as the volume

$$V = \left(\frac{\partial G}{\partial p}\right)_T \tag{10.2}$$

is discontinuous at the transition. Lines across which a discontinuity in density occurs are shown in the phase diagram, Fig. 10.1., while Fig. 10.2. shows schematically p as a function of V on isothermal lines for a system such as carbon dioxide. By varying the thermodynamic parameters, however, the liquid and gas phases can be made more similar until they can no longer be distinguished at a critical point c, shown in Figs. 10.1 and 10.2. On the isothermal $T = T_c$ there is no discontinuity in density, but we see from Fig. 10.2 that $(\partial p/\partial V)_T$ is zero at the critical point, and thus the isothermal compressibility

$$\kappa = -\frac{1}{V}\left(\frac{\partial V}{\partial p}\right)_T = -\frac{1}{V}\left(\frac{\partial^2 G}{\partial p^2}\right)_T \tag{10.3}$$

diverges at the critical point. The transition is then said to be second order, or continuous. The term critical point may not always have the right implication. Fig. 10.3 shows a sketch of the phase diagram of ^{4}He in the low temperature region. There is an entire line of critical points called the λ-line, across which the transition from fluid to superfluid is continuous.

Other examples of systems in which critical points (or lines) occur are paramagnet ⟷ ferromagnet, and disordered ⟷ ordered structure in a binary alloy such as CuZn. In a second order magnetic phase transition the spontaneous magnetization M has no discontinuity, and M is zero above T_c, while $(\partial M/\partial T)$

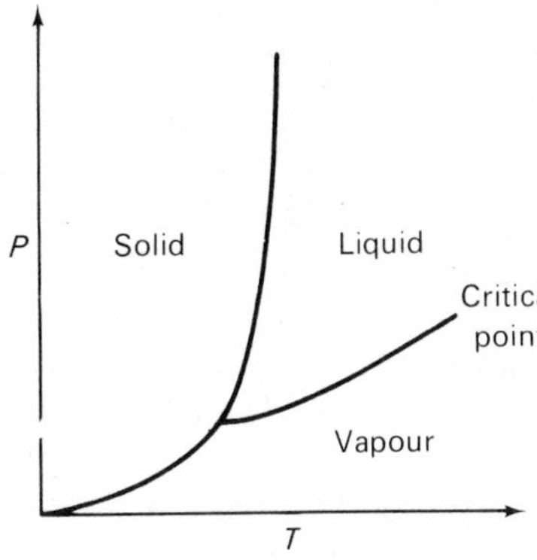

Figure 10.1 Phase diagram in the pressure–temperature plane. (After Heller, 1967, *Rep. Prog. Phys.* **30**, 731.)

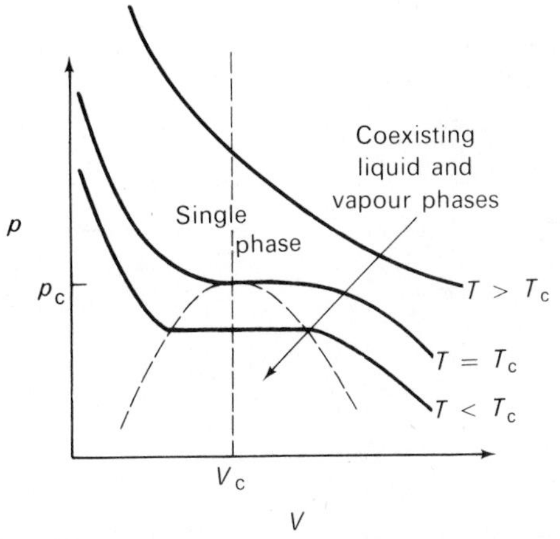

Figure 10.2 Isothermal lines in the pressure–volume plane, for a fluid. (After Heller, 1967, *Rep. Prog. Phys.* **30**, 731.)

diverges at T_c, as shown in Fig. 10.4b. If the material has a unique direction of easy magnetization, M may have either sense in this direction. The magnetic susceptibility $(\partial M/\partial H)_T$ diverges at the critical temperature, as shown in Fig. 10.4d. The magnetization M is referred to as the order parameter for the transition. The analogous quantity for the gas–liquid phase transition is not

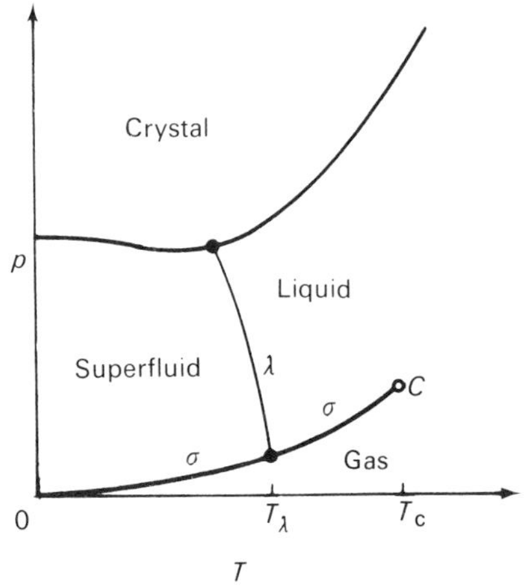

Figure 10.3 Phase diagram of helium at low temperatures. (After Heller, 1967, *Rep. Prog. Phys.* **30**, 731.)

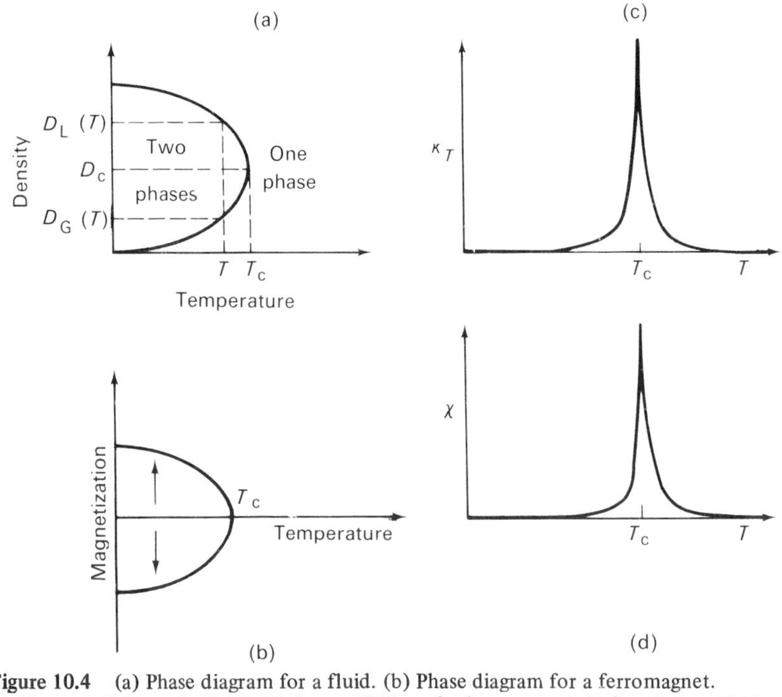

Figure 10.4 (a) Phase diagram for a fluid. (b) Phase diagram for a ferromagnet. (c) Compressibility of a fluid. (d) Susceptibility of a ferromagnet. (After Heller, 1967, *Rep. Prog. Phys.* **30**, 731.)

immediately apparent, but is in fact the difference in density between liquid and gas, $D_l - D_g$, on the co-existence curve, as shown in Fig. 10.4a. The compressibility of the fluid (Fig. 10.4c) is the analogue of the magnetic susceptibility.

The appropriateness of the term 'order parameter' is most apparent in considering an alloy such as CuZn. Above $T_c \triangleq 730$ K, X-ray reflections from this alloy are characteristic of a body-centred cubic structure. Evidently Cu and Zn atoms are distributed at random on the two interpenetrating simple cubic lattices which together make up the b.c.c. lattice, as illustrated in two dimensions in Fig. 10.5a. Below T_c there is a preponderance of Cu atoms on

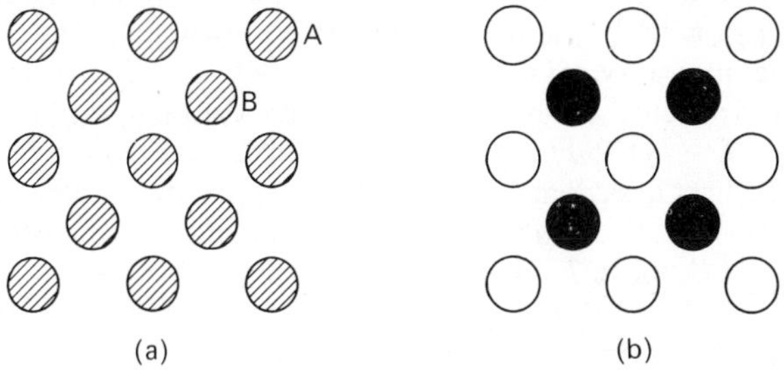

Figure 10.5 (a) Above T_c, Cu and Zn atoms are randomly distributed over A and B sites. (b) Below T_c, Cu atoms (open circles) order on to A sites.

one sub-lattice, denoted A. Complete order, illustrated in Fig. 10.5b, is in principle† attained at 0 K. The order parameter is defined by

$$\eta = \frac{2N_A - N}{N} \tag{10.4}$$

where N_A is the number of Cu atoms on A sites and N the total number of Cu atoms. The variation of η with T is similar to that of M in Fig. 10.4b. The order parameter can be measured experimentally from the intensities of those X-ray reflections (for which the sum of the indices $h_1 + h_2 + h_3$ is odd) which appear when the structure becomes simple cubic, as it is when A and B sites are no longer equivalent. For this system there is no physical quantity which corresponds to the susceptibility. Although η is zero above T_c, one can define a short range order parameter in terms of the probability that an atom has an excess of near neighbours of the other species, and this remains non-zero well above T_c. Corresponding short range order effects are often strongly evident above magnetic phase changes.

† In practice, the diffusion processes required for atoms to go from 'wrong' sites to 'right' ones become extremely slow at and below room temperature.

A notable feature of second order phase transitions is the large fluctuations which occur close to the critical point. In a liquid the density fluctuates, in a magnetic crystal the magnetization, leading to a high intensity of scattering of radiation. This is the phenomenon known as critical opalescence or critical scattering and is evidently associated with the divergence of the compressibility or susceptibility. A subject of great interest to both experimentalists and theoreticians in recent years has been the investigation of the behaviour of the system in the 'critical region' in which some of its properties are to a considerable extent determined by the existence of fluctuations. Fluids and magnetic systems have been found to be remarkably similar in their behaviour. In the former the density difference $D_l - D_g$, and in the latter the magnetization M, vary as $(T_c - T)^{\beta}$, with β close to 1/3 for a range of materials. The compressibility κ and susceptibility χ vary as $(T - T_c)^{-\gamma}$ with γ close to 4/3 in many instances Other 'critical exponents' can be defined, such as α for the specific heat C_V, which varies as $(T - T_c)^{-\alpha}$. For several systems α has a value of order 1/10 so that the specific heat has anomalous values in a comparatively narrow range of temperature above T_c. Corresponding exponents γ' and α' give the variation of χ (or κ) and of C_V when T is below T_c, for example

$$C_V \sim (T_c - T)^{-\alpha'} \tag{10.5}$$

It is usually the case that $\gamma' \doteq \gamma$. Simple functional relations such as Equ. 10.5 cannot be expected to apply over a wide range of temperature and those which have been quoted are to be regarded as limiting forms valid in the critical region.

10.2 Displacive phase transitions

By a displacive phase transition we mean a transition from one (ordered) crystal structure to another, in which each atom moves relative to its near neighbours by an amount small compared with a unit cell dimension. If the transition is second order the symmetry of the crystal is lowered as the temperature decreases through T_c, but without any abrupt discontinuity. A displacive transition is to be distinguished from a reconstructive one, from one crystal structure to a completely different one, such as occurs in certain alkali halides which make a first order transition from the NaCl to the CsCl type of structure. A displacive phase transition is illustrated in two dimensions in Fig. 10.6. Fig. 10.6a shows a diatomic crystal with a square unit cell, the two-dimensional analogue of the CsCl type of structure. Above T_c atoms of type 1 occupy the position (0, 0), those of type 2 the position $(\frac{1}{2}, \frac{1}{2})$. Below T_c atoms of type 2 are displaced by an amount δ relative to those of type 1, as shown, in Fig. 10.6b. The order parameter η can be taken to be δ, and the symmetry changes from square to rectangular as T decreases through T_c. It should be emphasized that while each atom of type 2 will not at any instant be displaced by exactly the same amount from $(\frac{1}{2}, \frac{1}{2})$ because of thermal vibration, the value of δ is the same in every unit cell when the position is averaged over a time somewhat greater than 10^{-12} sec, which is the reciprocal of a typical phonon frequency. Thus both phases have completely ordered structures, a feature which distinguishes a displacive structural phase transition from others which were briefly discussed in §10.1.

We now consider the three-dimensional analogue of Fig. 10.6a, that is the CsCl type of structure, and express the displacements of the atoms in terms of normal coordinates, as discussed in Chapter 4. The generalization of Equ. 4.10 to a crystal with several atoms per primitive unit cell, but with each atom at a centre of symmetry, is that the displacement is given by

$$\mathbf{u}_{l\kappa} = (Nm_\kappa)^{-1/2} \sum_{\mathbf{q}j} B_j(\mathbf{q}) e_{\kappa j}(\mathbf{q}) \cos[\mathbf{q}\cdot\mathbf{r}_{l\kappa} - \omega_j(\mathbf{q})t + \alpha_j(\mathbf{q})] \qquad (10.6)$$

As before the subscripts l and κ identify particular unit cells and types of atom respectively, we could have $\kappa = 1$ for Cs and $\kappa = 2$ for Cl in a diatomic

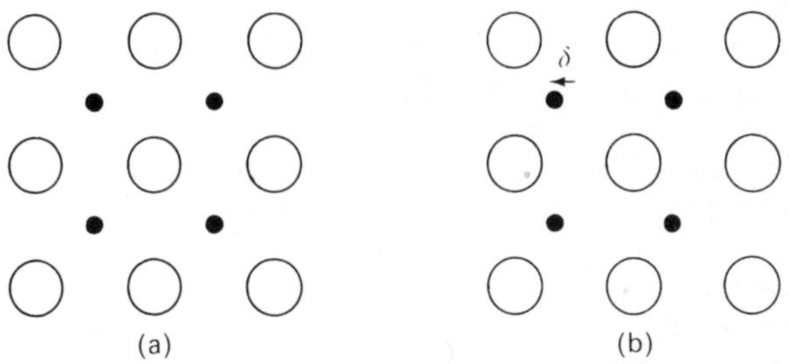

Figure 10.6 (a) A two-dimensional diatomic structure above T_C. Atoms are situated on four-fold axes. (b) Below T_C the relative displacement δ removes the four-fold symmetry.

crystal, for example. The equilibrium position of an atom is $\mathbf{r}_{l\kappa} = \mathbf{r}_l + \mathbf{r}_\kappa$. The frequency of a mode $\omega_j(\mathbf{q})$ and the eigenvectors $\mathbf{e}_{\kappa j}(\mathbf{q})$ are determined by the interatomic force constants, as discussed in Chapter 4, although for particular values of $\mathbf{q}$ and particular branches j of the dispersion curves, the $\mathbf{e}_{\kappa j}(\mathbf{q})$ may be determined by symmetry. For a crystal in equilibrium at temperature T the mean square value of the amplitude $B_j(\mathbf{q})$ is determined by Equ. 4.12. The displacement of any atom due to one particular mode $\mathbf{q}'j'$ depends on $N^{-1/2}$ and is therefore negligibly small. However this is no longer the case when the frequency of a particular mode $\omega_{j'}(\mathbf{q}')$ tends to zero. The right hand side of Equ. 4.12 then tends to the value $k_B T$, and the amplitude can be arbitrarily large. Also, for $\omega_{j'}(\mathbf{q}')$ just zero the displacement produced by this mode is not oscillatory, but static. The atomic displacements would in this situation be limited by the anharmonic character of the interatomic potential. We see from Equ. 10.6 that the condition for the displacement to be the same in every unit cell, i.e. for $\mathbf{u}_{l\kappa}$

to be independent of l, as in Fig. 10.6b, is that the unstable mode should have $q' = 0$. In this and other instances where $\mathbf{q}'$ is at a special position in the Brillouin zone, several modes have the same frequency but different eigenvectors, and can become unstable simultaneously. In the harmonic approximation the displacement resulting from the instability can be in various directions and which actually develops will depend on anharmonic effects. Domains in which the displacement is in symmetry related directions may be formed in different parts of the crystal.

A mechanism by which an instability can be brought about in a mode for which $q = 0$, while leaving other modes stable, is readily demonstrated for an ionic crystal. For simplicity we use a rigid ion model although the discussion can be easily extended to take account of electronic polarizability. The frequency of a transverse optic mode for which $q = 0$ is given by Equ. 7.25.

$$m\omega_{\mathrm{T}}^2(0) = \beta - \frac{(Ze)^2}{3v\epsilon_0} \tag{10.7}$$

The subscript $j = T$ was used to identify a transverse optic mode. We see that in principle a zero frequency can result from a cancellation of forces originating in the short range and Coulomb interactions respectively. From Equ. 7.27 we see that the longitudinal optic mode having $q = 0$ remains stable, and it is possible to choose force constants such that all other modes remain stable, although those for which $\mathbf{q}$ is close to zero in transverse optic branches will have low frequencies.

In the harmonic approximation $\omega_{\mathrm{j}'}(\mathbf{q}')$ is independent of temperature, and a zero frequency would remain zero, or if $\omega_{\mathrm{j}'}^2(\mathbf{q}')$ were less than zero for a mode of the cubic structure, a structure of lower symmetry would be the stable phase at all temperatures. We saw in Chapter 8, however, that anharmonic interactions make the effective frequency of any mode a function of temperature and of the frequency with which the crystal is probed. As an approximation we can ignore this latter complication and introduce the quasi-harmonic frequency $\tilde{\omega}_{\mathrm{j}}(\mathbf{q})$, which is the frequency determined experimentally by neutron spectroscopy for example, and which depends only on temperature. We shall take the stability of the crystal to be determined by the values of $\tilde{\omega}_{\mathrm{j}}(\mathbf{q})$. It is then possible to have $\omega_{\mathrm{j}'}^2(\mathbf{q}') < 0$, but $\tilde{\omega}_{\mathrm{j}'}^2(\mathbf{q}') > 0$. The temperature dependence of the shift in frequency is very complicated, as we saw in §8.4, but at relatively high temperatures the difference between $\tilde{\omega}_{\mathrm{j}}^2(\mathbf{q})$ and $\omega_{\mathrm{j}}^2(\mathbf{q})$ is predicted to be proportional to temperature. We may therefore postulate that

$$\tilde{\omega}_{\mathrm{j}'}^2(\mathbf{q}') = B_{\mathrm{j}'}(\mathbf{q}')(T - T_{\mathrm{c}}) \tag{10.8}$$

where T_{c} is the lower limit of the stability of the cubic structure and $B_{\mathrm{j}'}(\mathbf{q}')$ is a constant. Modes other than $\mathbf{q}'j'$ are assumed to remain stable.

This postulate finds immediate support in the properties of certain ferroelectric crystals such as $BaTiO_3$, the structure of which is illustrated in Fig. 10.7, for a temperature above 130°C. A ferroelectric transition is a special case of a structural phase transition (not necessarily displacive) in which the change of symmetry is such that the crystal develops a spontaneous polarization below T_c (ref. 10.1). If the transition to the ferroelectric phase is displacive the mode involved must be a transverse optic mode of wavenumber zero. In fact the structure shown in Fig. 10.6b would exhibit a spontaneous polarization, since each unit cell has the same dipole moment $Ze\delta$ if we take the atoms to be rigid

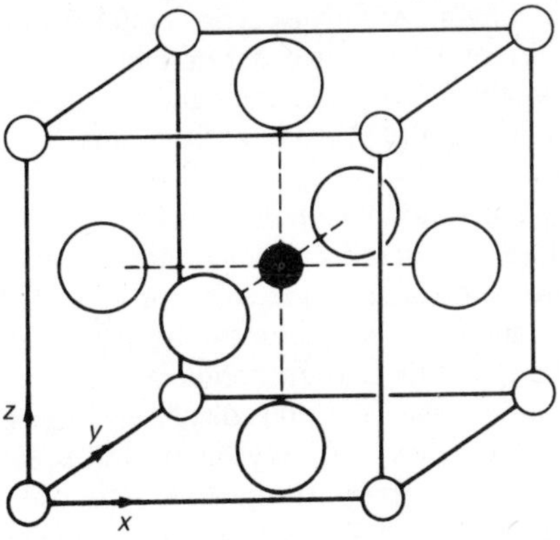

Figure 10.7 The structure of $BaTiO_3$ in the cubic phase. Ba is at a cube corner, Ti in the body centre, and O on face centres. The six oxygens shown are at the corners of a regular octahedron.

ions of charge $\pm Ze$. The static dielectric constant of a cubic crystal is given in the quasi-harmonic approximation by

$$\epsilon(0) = \epsilon(\infty) + \sum_j \frac{A_j}{\tilde{\omega}_j^2(0)} \tag{10.9}$$

(compare Equ. 8.11) where each A_j is a constant. The sum is over all transverse optic modes for which $q = 0$, of which there are four in $BaTiO_3$, although it can be shown that for one of these the constant A_j is zero. Thus equations 10.9 and 10.8 taken together predict that for $T > T_c$,

$$\epsilon(0) = \epsilon(\infty) + \epsilon_1(0) + \frac{\mathscr{C}}{T - T_c} \tag{10.10}$$

where $\mathscr{C}$ is a constant, and $\epsilon_1(0)$ is the contribution to the dielectric constant of the two transverse optic modes which are assumed not to have an anomalous temperature dependence. Equ. 10.10 is just what is found experimentally (ref. 10.1), with the qualification that for $BaTiO_3$ the transition is first order with the transition temperature T_0 a few degrees above the temperature T_c which fits Equ. 10.10. The transition may be described as 'nearly second order' in that the dielectric constant rises to a value of several thousand but is still finite before the transition takes place, with a discontinuity in $\epsilon(0)$. The thermodynamic theory (ref. 10.1) shows that if the crystal could be constrained to be at constant volume rather than constant pressure, the transition would be second order. What distinguishes ferroelectric crystals from others which make structural phase transitions is that the order parameter (the spontaneous polarization) is comparatively easy to detect and measure, and the susceptibility (which is simply $\epsilon(0) - 1$) is a physical quantity whose divergence at T_c can be directly demonstrated.

The spontaneous polarization which develops below 130°C in $BaTiO_3$ is parallel to a cube axis, and in practice domains are formed with different directions of spontaneous polarization in adjacent parts of the crystal. In an applied field the spontaneous polarization can be aligned along one of the three axes. The crystal structure at room temperature is tetragonal. Taking the Ba atom to be at the origin, the Ti atom is displaced from $\frac{1}{2}, \frac{1}{2}, \frac{1}{2}$ by 0·07 Å in the direction of the polarization, with oxygen atoms displaced by approximately the same amount from face-centre positions in the opposite direction. Since the unit cell dimension is $a = 4{\cdot}00$ Å, these relatively small displacements tend to confirm that the phase transition is displacive.

It may however be difficult, both in practice and in principle, to distinguish with certainly between a displacive transition and one involving an ordering process of a less drastic kind than we considered in § 10.1. This is illustrated in Fig. 10.8, which shows a two-dimensional structure in which atoms of type 1 are ordered, but above T_c each atom of type 2 is located with equal probability on one of a group of four possible sites, each a distance Δ from the centre of the cell, giving a structure with square symmetry. Below T_c one site, say the one furthest to the left, is occupied with a probability $p^* > \frac{1}{4}$, the others having equal probability $\frac{1}{3}(1 - p^*)$. The symmetry of the structure is then rectangular and, provided that Δ is small compared with the unit cell dimension a, it would not be possible by means of macroscopic measurements to distinguish this situation from that illustrated in Fig. 10.6b. We may define the order parameter as

$$\eta = \tfrac{1}{3}(4p^* - 1)\Delta \qquad (10.11)$$

the value of which lies between 0 and Δ, and is proportional to the spontaneous polarization. However, if Δ is a few times larger than the mean amplitude of thermal vibration, the situation corresponding to Figs. 10.6a and 10.8 could be distinguished by a crystal structure determination, and the fact that each individual atom of type 2 did not have an environment of cubic symmetry in the three-dimensional structure corresponding to Fig. 10.8 could be detected by a technique such as nuclear magnetic resonance, provided that a particular off-centre position was maintained for a time of order 10^{-7} sec. One must remember that each atom moves in a potential minimum, but when the potential barrier between the various off-centre positions is smaller than $k_B T$ it will pass between these positions in a time which is of the same order of magnitude as the period of a typical phonon, and any possibility of distinguishing the two types of

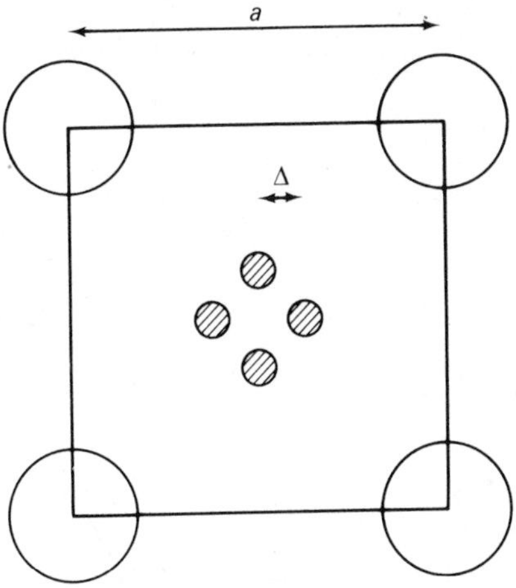

Figure 10.8 One unit cell of a square structure in which an atom of one type occupies randomly one of four possible sites (shaded) each a distance Δ from the centre of the cell.

transition is apparently lost. It may be however that the distinction between them is then illusory in that the dynamics of the 'disordered' structure can be treated in the same way as for an ordered structure with anharmonic effects being somewhat more important.

Structural phase transitions in which ordering of atoms undoubtedly occurs with $\Delta \ll a$, are well known. For example the structure of KH_2PO_4 above $T_c = 120$ K has each hydrogen atom statistically distributed over two sites separated by 0·4 Å. Below T_c the hydrogens order preferentially on to one of those sites, with other atoms making smaller displacements. While the dynamics of such transitions are of considerable interest, we shall confine our attention to displacive transitions, as defined above.

10.3 The dynamics of soft modes

We saw in the previous section that the presence of an optic mode which has a low and strongly temperature dependent frequency can be inferred from the behaviour of the dielectric constant of a ferroelectric crystal, particularly when considered in conjunction with the change of crystal structure which occurs at the phase transition. The existence of such modes can also be demonstrated more directly by infra-red spectroscopy and by neutron inelastic scattering. They are often referred to as soft modes, a piece of jargon which has the merit of brevity.

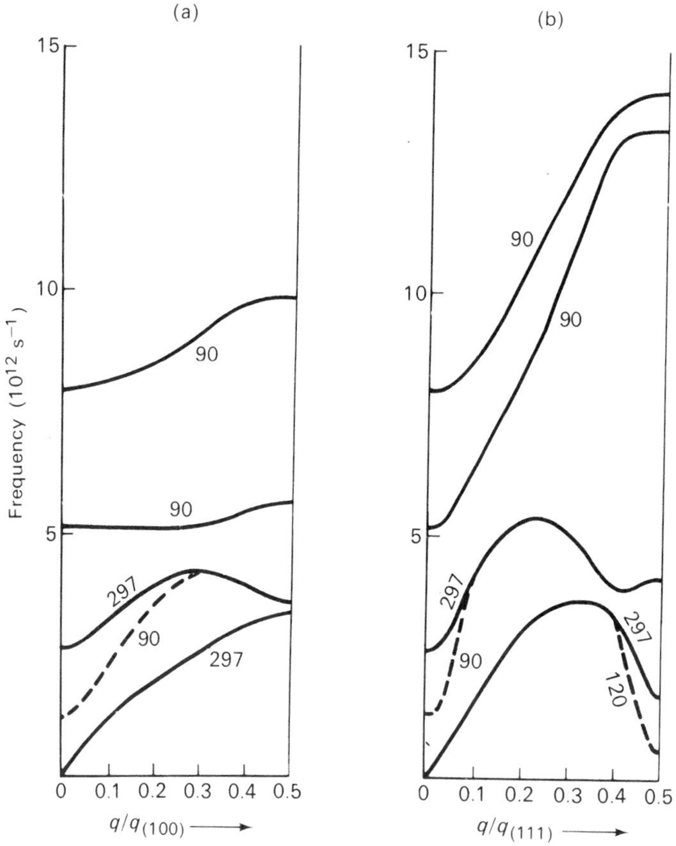

Figure 10.9 Experimentally measured frequencies $\bar{\omega}_j(\mathbf{q})/2\pi$ for $SrTiO_3$. In (a) **q** is parallel to [100], in (b) to [111]. Only transverse modes are shown. A number such as 90 indicates the measurements were made at 90 K; where frequencies vary considerably with temperature a branch at a lower temperature is shown by a dotted line. (After Stirling, 1972, *J. Phys. C*, **5**, 2711.)

$SrTiO_3$ has been the subject of a number of investigations. At room temperature the crystal structure is the perovskite structure shown in Fig. 10.7. The phonon dispersion curves have been determined by neutron scattering (ref. 10.2) and are shown in Fig. 10.9 for **q** parallel to [100] and to [111] directions. There are five atoms per unit cell, so that for a general value of **q** there are fifteen branches, i.e. values of *j*. However for **q** in the directions just mentioned, pairs of transverse branches have the same frequency giving in effect five transverse and five longitudinal branches, not all of which are shown

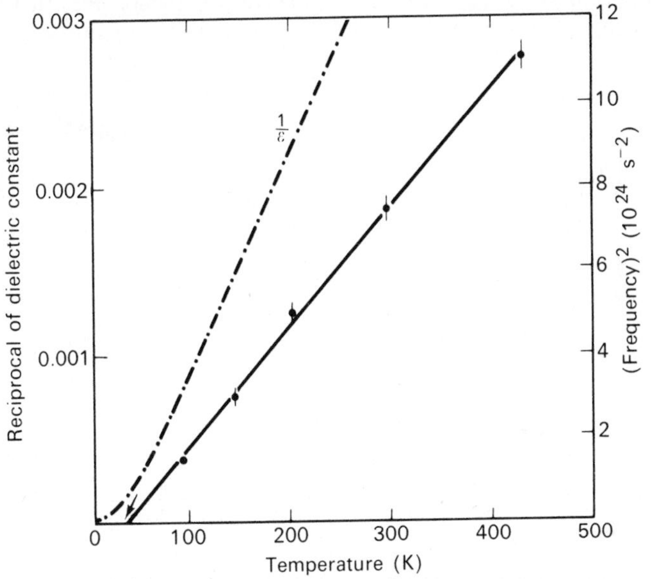

Figure 10.10 The square of the frequency of a transverse optic mode in $SrTiO_3$ as a function of temperature. The dotted line gives the reciprocal of the static dielectric constant. (After Cowley, 1962, *Phys. Rev. Lett.* 9, 159.)

in Fig. 10.9 since the highest frequencies are difficult to measure. It will be noticed that for values of q close to the origin there is a dip in the frequency of the lowest transverse optic branch, at room temperature. Fig. 10.10 shows $\omega_{j'}^2(\mathbf{q}')$ for the mode in this branch having $q' = 0$, as a function of temperature. The variation is quite accurately linear over the range shown. The same diagram shows $1/\epsilon(0)$, which is also found to vary as $T - T_c$ in this range, with the same value $T_c \triangleq 30$ K as fits Equ. 10.8. However, the crystal does not make a transition to a ferroelectric phase; below about 60 K the dielectric constant increases less rapidly than predicted by Equ. 10.10 and tends to a constant value as 0 K is approached. Evidently the mode we are considering remains stable, but only just so, at 0 K, and this has been confirmed by measurements of its frequency at low temperatures, shown in Fig. 10.11.

Accurate measurements of the unit cell dimensions of $SrTiO_3$ showed that the crystal becomes tetragonal below a temperature $T_c' = 108$ K. The occurence of a second order phase transition is confirmed by the variation of the specific heat, but the dielectric constant apparently varies quite smoothly through this temperature. This was for some time a puzzle, until it was found that this is an example of a displacive structural phase transition for which the mode involved is not at the Brillouin zone centre, but at the zone boundary. Fig. 10.12 shows in more detail than in Fig. 10.9 the lowest transverse acoustic branch for **q**

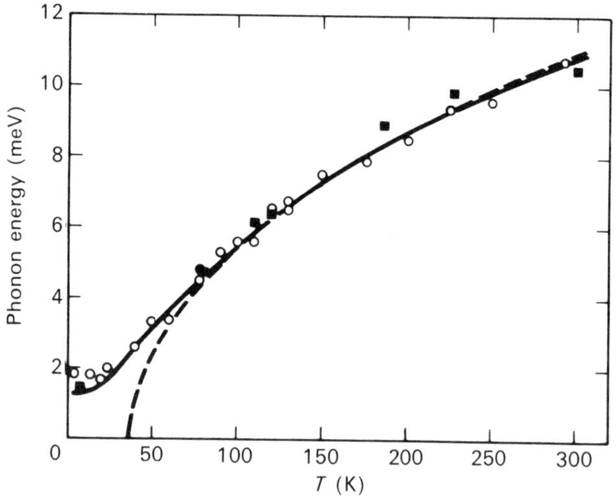

Figure 10.11 The frequency (given as $\hbar\omega_j(\mathbf{q})$) of a transverse optic mode in $SrTiO_3$ measured down to 4 K. The dashed line shows what the frequency would be if it continued to vary as $(T - T_c)^{1/2}$. (After Yamada and Shirane, 1969, *J. Phys. Soc. Japan* **26**, 396.)

parallel to [111]. The boundary of the Brillouin zone is at $\mathbf{q} = (2\pi/a)\,(\frac{1}{2}\frac{1}{2}\frac{1}{2})$, and as this point is approached there is a sharp dip in the dispersion curve. The square of the frequency, for $\mathbf{q}'$ at the zone boundary, is shown in Fig. 10.13. It is quite accurately linear and the frequency remains well defined at least to within a few degrees of the phase transition. Substitution of $\mathbf{q} = (2\pi/a)\,(\frac{1}{2}\frac{1}{2}\frac{1}{2})$ in Equ. 10.6 shows that the direction of displacement of equivalent atoms in adjacent unit cells is reversed. The actual change of structure which occurs in $SrTiO_3$ is illustrated in Fig. 10.14. One set of oxygen atoms, and the Ti and Sr atoms, are not displaced. The corresponding eigenvectors $\mathbf{e}_{\kappa j'}(\mathbf{q}')$ in Equ. 10.6 are zero. The axis through the undisplaced oxygens defines a unique axis, the z axis, which is vertical in Fig. 10.14. The remaining two sets of oxygen atoms are displaced as shown in this diagram, those in the layer above or in the layer below that shown being displaced in the opposite direction. In effect each regular octahedron which has Ti at its centre and O at each of its six corners is rotated about the z axis with the sense of the rotation reversed in adjacent unit cells. The order parameter for the transition is the angle ϕ shown in Fig. 10.14.

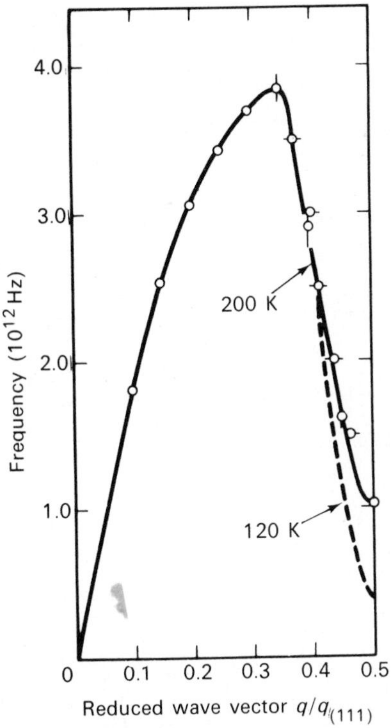

Figure 10.12 Frequencies in an acoustic branch of the phonon spectrum for $SrTiO_3$. The wavevector is parallel to [111]; a low and temperature dependent frequency is found at the boundary of the Brillouin zone. (After Cowley, Buyers and Dolling, 1969, *Solid State Commun.* 7, 181.)

Below T_c the frequencies of the modes which oscillate the octahedra about the x or y axes remain equal. This frequency increases rather slowly again as T decreases below T_c. The frequency of the mode which oscillates the octahedra about the z axis in the low temperature phase is higher, and varies more rapidly with T, as shown in Fig. 10.13. These modes have also been investigated by Raman spectroscopy.

A structural phase transition involving modes for which $\mathbf{q}' = (2\pi/a)(\frac{1}{2}\frac{1}{2}\frac{1}{2})$ also occurs in $LaAlO_3$. In the cubic phase oscillations of the octahedra about the

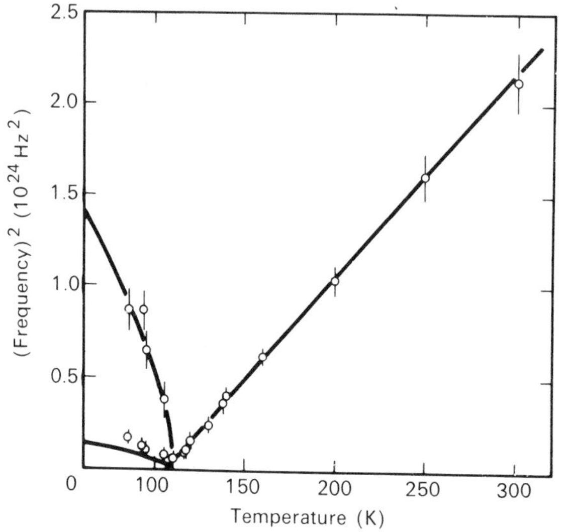

Figure 10.13 The square of the frequency of the mode for which the wavevector **q**′ is at the zone boundary (see Fig. 10.12), as a function of temperature. (After Cowley, Buyers and Dolling, 1969, *Solid State Commun.* 7, 181.)

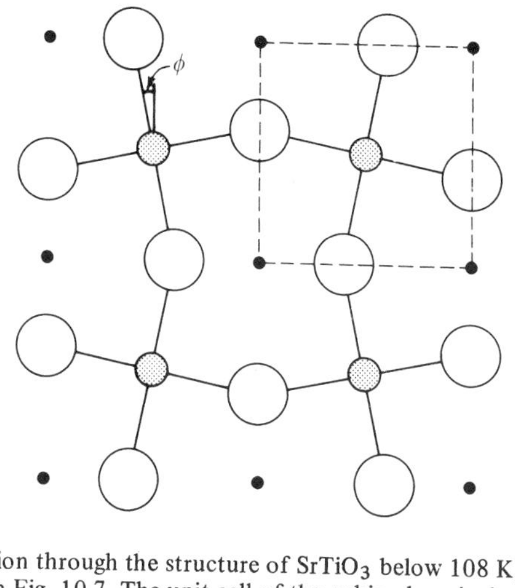

Figure 10.14 A section through the structure of $SrTiO_3$ below 108 K, at a level which corresponds to $z = \frac{1}{2}$ in Fig. 10.7. The unit cell of the cubic phase is shown by dotted lines, and the z axis is now perpendicular to the plane of the paper.

three cubic axes have of course the same frequency, although in $SrTiO_3$ a rotation about only one of them develops below T_c. In $LaAlO_3$ there are equal rotations about all three axes to give a resultant rotation about a cube diagonal below T_c. Which of several possible structures develops in the low temperature phase depends on anharmonic effects, as we already commented in § 10.2.

The lattice dynamics of $BaTiO_3$, the supposed prototype for displacive ferroelectric transitions, has proved to be somewhat more complicated than was anticipated. In the cubic phase when **q** is in the [100] direction for example, the neutron spectrum $S_1(\mathbf{K}\omega)$ shows considerable intensity at low frequencies, not only for q close to zero but for all values of **q** to the zone boundary at $(\pi/a)(1\ 0\ 0)$. Evidently the lowest transverse optic branch has a low frequency not only for $q = 0$. There is, however, for a particular value of **q**, no peak in the spectrum which would define a quasi-harmonic frequency $\tilde{\omega}_j(\mathbf{q})$. To understand this we must consider again the form of $S_1(\mathbf{K}\omega)$. For a soft mode, we can replace $(\bar{n}_j(\mathbf{q}) + \frac{1}{2})\,\hbar\,\omega_j(\mathbf{q})$ by k_BT. In the situation where the damping constant of a mode is relatively small, we have from Equ. 8.14 that

$$S_1(\mathbf{K}\omega) \propto \frac{k_BT}{\tilde{\omega}_j^2(\mathbf{q})}\left\{\frac{\tilde{\Gamma}_j(\mathbf{q})}{(\omega - \tilde{\omega}_j(\mathbf{q}))^2 + \tilde{\Gamma}_j^2(\mathbf{q})} + \frac{\tilde{\Gamma}_j(\mathbf{q})}{(\omega + \tilde{\omega}_j(\mathbf{q}))^2 + \tilde{\Gamma}_j^2(\mathbf{q})}\right\} \tag{10.12}$$

In the limit of zero damping, $\tilde{\omega}_j(\mathbf{q})$ reverts to $\omega_j(\mathbf{q})$ and $\tilde{\Gamma}_j(\mathbf{q})$ becomes zero, so that the two Lorentzian peaks, corresponding to neutron scattering with energy loss and energy gain respectively, become δ-functions, in agreement with Equ. 5.13. However for larger values of the damping constant we have to use the fact, mentioned in § 8.2, that $S_1(\mathbf{K}\omega)$ is determined by the imaginary part of the response function of a mode. This is given in the quasi-harmonic approximation by Equ. 8.21, leading to the result that

$$S_1(\mathbf{K}\omega) \propto \frac{4\tilde{\Gamma}_j(\mathbf{q})k_BT}{(\omega^2 - \tilde{\omega}_j^2(\mathbf{q}))^2 + 4\omega^2\tilde{\Gamma}_j^2(\mathbf{q})} \tag{10.13}$$

For $\tilde{\Gamma}_j(\mathbf{q}) \ll \tilde{\omega}_j(\mathbf{q})$ this is no different from the spectrum given by Equ. 10.12. However for $2\tilde{\Gamma}_j^2(\mathbf{q}) \geqslant \tilde{\omega}_j^2(\mathbf{q})$ Equ. 10.13 shows that the spectrum has no maximum at or near $\pm\,\tilde{\omega}_j(\mathbf{q})$, but only at $\omega = 0$. The mode is then said to be overdamped. Evidently this is the case for certain modes in the lowest transverse optic branch of $BaTiO_3$. Whether this is the whole story is not entirely clear. There is evidence which favours the view that in $BaTiO_3$ the situation verges on that discussed at the end of § 10.2, with Ti and O atoms moving more easily parallel to the cube axes than in other directions.

We may note that as a displacive transition is approached and $\tilde{\omega}_{j'}(\mathbf{q}')$ tends to zero above T_c, this mode, and those having adjacent values of **q**, which necessarily also have low frequencies, build up regions in the crystal in which the structure briefly is that of the low temperature phase. Radiation scattered by the soft modes is more nearly elastically scattered as $\tilde{\omega}_{j'}(\mathbf{q}')$ decreases. The intensity integrated over frequency, $\int S_1(\mathbf{K}\omega)d\omega/2\pi$, can be shown from Equ. 10.13 to be independent of $\tilde{\Gamma}_{j'}(\mathbf{q}')$ and proportional to $k_B T/\tilde{\omega}^2_{j'}(\mathbf{q}')$. It therefore diverges as T approaches T_c. This description of critical fluctuations and of critical scattering is adequate provided T is not too close to T_c, but precisely in the critical region the quasi-harmonic approximation cannot be expected to be valid. Recent experimental work (ref. 10.3) has indeed shown that the above description is an oversimplification.

10.4 General theories of phase transitions

Only a very brief outline of this topic can be given here; for an excellent account the reader is referred to a book by Stanley, (ref. 10.4). The models used by theoreticians in investigating phase transitions are usually conceptually simple but involve formidable mathematical problems. The Ising model is a simple and not unrealistic model for a magnetic phase transition. Magnetic dipoles which can be oriented in either sense in a unique direction occupy the points of a lattice. It is conventional to have $S_l = +1$ denote the up orientation and $S_l = -1$ the down orientation, although it is a spin-$\frac{1}{2}$ system which can assume only two orientations. The Hamiltonian of the system is taken to be

$$\mathcal{H} = -\tfrac{1}{2} \sum_{ll'} J_{ll'} S_l S_{l'} \tag{10.14}$$

Notice that this model has no dynamics in that no mechanism to change S_l is postulated, and therefore only equilibrium properties can be calculated. We shall assume the interaction $J_{ll'}$ is zero except between nearest neighbours, when $J_{ll'} > 0$. Clearly at 0 K all spins will point in one direction, giving complete order. At a higher temperature the minimum free energy will be given by configurations having greater energy but increased entropy of disorder. The problem is to find the number of configurations having the same energy, the free energy, and the degree of correlation between spins separated by a given amount. It was shown by Ising that in one dimension (a linear chain of spins), long range order, that is spontaneous magnetization, is complete at 0 K but zero for all other temperatures! Short range order, that is correlation between neighbouring spins, is found, but there is no phase transition, although we might put $T_c = 0$. By a mathematical tour de force Onsager solved the problem of the two-dimensional Ising model having a rectangular lattice, with nearest neighbour interactions. Taking this to be J to all four neighbours of a particular spin, Onsager found that a phase transition occurred when

$$k_B T_c = 2{\cdot}268J \tag{10.15}$$

At this temperature the long range order falls steeply but continuously to zero. The problem of the three-dimensional Ising model has not been solved, but approximate methods have given the critical exponents, defined in §10.1. For α, β and γ they are respectively $\frac{1}{8}$, $\frac{5}{16}$ and $\frac{5}{4}$ when the lattice is simple cubic, values not greatly different from those which apply to actual phase transitions. It has been found possible to adapt the Ising model so that it applies also to the liquid $\longleftrightarrow$ gas transition and the order $\longleftrightarrow$ disorder transition in a binary alloy.

In the mean field or molecular field theory of ferromagnetism introduced by Weiss, each magnetic dipole is taken to be acted on by a field proportional to the spontaneous magnetization. In terms of the Ising model, the spins which are nearest neighbours of a particular spin are replaced by their average value over the whole crystal. For the two-dimensional Ising model this gives a phase transition at a temperature T_M given by

$$k_B T_M = 4J \qquad (10.16)$$

which is higher than the correct solution given by Equ. 10.15. The critical exponents predicted by the mean field theory are $\alpha = 0$ (i.e. the specific heat does not diverge), $\beta = \frac{1}{2}$ and $\gamma = 1$. Recently it has been shown that the exact solution for the Ising model tends to that given by the mean field approximation as the interaction $J_{ll'}$ extends to more distant neighbours; that is, in the limit of infinitely long range (but infinitely weak!) interactions, the mean field approximation is exact. Somewhat surprisingly, the assumptions made by van der Waals in deriving his famous equation of state are equivalent to the mean field approximation and lead to the classical values for α, β and γ of 0, $\frac{1}{2}$ and 1. These of course do not agree with the values which are found for actual gas $\longleftrightarrow$ liquid phase transitions, which we quoted in §10.1

There is no obvious connection between these approaches and that which we outlined earlier for displacive structural transitions, which as we have seen is based on a consideration of dynamical properties and mechanical stability. The latter approach can however be shown to be closely related to a general theory of second order phase transitions proposed by Landau (ref. 10.5). Landau's theory places considerable emphasis on the change of symmetry which occurs at the transition. He comes to the conclusion that the difference between the free energies of the two phases can be expressed as a power series in the order parameter,

$$G = \tfrac{1}{2}A\eta^2 + \tfrac{1}{4}B\eta^4 + \tfrac{1}{6}C\eta^6 + \cdots \qquad (10.17)$$

For simplicity we are limiting consideration to the situation where there is a unique direction along which the magnetization or polarization, etc. develops. In Equ. 10.17, A is taken to vary as $T - T_c$, with B and C constant. For stability $(\partial G/\partial\eta) = 0$ and $(\partial^2 G/\partial\eta^2) \geqslant 0$. This leads to the result that η is finite below T_c, with a critical exponent $\beta = \frac{1}{2}$. Taking A to vary as $T - T_c$ corresponds to setting $\gamma = 1$. The connection with the quasi-harmonic theory of lattice dynamics is that A is found to be proportional to $\tilde{\omega}_{j'}^2(\mathrm{q}')$, and B and C can be related to the anharmonic potential.

We have at this point come within sight of the frontiers of knowledge of this particular topic and this therefore seems to be an appropriate point at which to end.

References and suggestions for further reading

10.1 Jona, F. and Shirane, G. 1962, *Ferroelectric Crystals*, Pergamon Press, Oxford.
10.2 Stirling, W. 1972, *J. Phys. C*, 5, 2711.
10.3 *Structural Phase Transitions and Soft Modes*, 1971, Ed. E. J. Samuelson, E. Anderson and J. Feder, Scandinavian University Books.
10.4 Stanley, H. E. *Introduction to Phase Transitions and Critical Phenomena*, 1971, Clarendon Press, Oxford.
10.5 Landau, L. D. and Lifshitz, E. M. 1959, *Statistical Physics*, Pergamon Press, Oxford.

Author Index

Bold numbers indicate main references

Subject Index